WHY IS SEX FUN?

WHY IS SEX FUN?

The Evolution of Human Sexuality

JARED DIAMOND

BASIC
BOOKS

A Member of the Perseus Books Group

The Science Masters Series is a global publishing venture consisting of original science books written by leading scientists and published by a worldwide team of twenty-six publishers assembled by John Brockman. The series was conceived by Anthony Cheetham of Orion Publishers and John Brockman of Brockman Inc., a New York literary agency, and developed in coordination with Basic Books.

• • • • • • • • • • • • •

The Science Masters name and marks are owned by and licensed to the publisher by Brockman Inc.

• • • • • • • • • • • • •

Library of Congress Cataloging-in-Publication Data

Diamond, Jared M.
    Why is sex fun? : the evolution of human sexuality / by Jared Diamond. — 1st ed.
        p. cm.
    Includes index.
    ISBN-10 0-465-03127-7 (cloth)
    ISBN-13 978-0-465-03127-6 (cloth)
    ISBN-10 0-465-03126-9 (paper)
    ISBN-13 978-0-465-03126-9 (paper)
    1. Sex. 2. Sex customs. I. Title.
HQ21.D48 1997
306.7—DC21   96-44065
                              CIP

EBC   07 08 09   27 26 25 24

To Marie,
My best friend, coparent, lover, and wife

CONTENTS

The subject of sex preoccupies us. It's the source of our most intense pleasures. Often it's also the cause of misery, much of which arises from built-in conflicts between the evolved roles of women and men.

This book is a speculative account of how human sexuality came to be the way it now is. Most of us don't realize how unusual human sexual practices are, compared to those of all other living animals. Scientists infer that the sex life of even our recent apelike ancestors was very different from ours today. Some distinctive evolutionary forces must have operated on our ancestors to make us different. What were those forces, and what really is so bizarre about us?

Understanding how our sexuality evolved is fascinating not only in its own right but also in order to understand our other distinctively human features. Those features include our culture, speech, parent-child relations, and mastery of complex tools. While paleontologists usually attribute the evolution of these features to our attainment of large brains and upright posture, I argue that our bizarre sexuality was equally essential for their evolution.

Among the unusual aspects of human sexuality that I discuss are female menopause, the role of men in human societies, having sex in private, often having sex for fun rather than for procreation, and the expansion of women's breasts even before use in lactation. To the layperson, these features all seem almost too natural to require explanation. On reflection, though, they prove surprisingly difficult to account for. I'll also discuss the function of men's penises and the reasons women but not men nurse their babies. The answers to these two questions seem utterly obvious. Within even these questions, though, lurk baffling unsolved problems.

Reading this book will not teach you new positions for enjoying intercourse, nor will it help you reduce the discomfort of menstrual cramps or menopause. It will not abolish the pain of discovering that your spouse is having an affair, neglecting your joint child, or neglecting you in favor of your child. But this book may help you understand why your body feels the way it does, and why your beloved is behaving the way he or she is. Perhaps, too, if you understand why you feel driven to some self-destructive sexual behavior, that understanding may help you to gain distance from your instincts and to deal more intelligently with them.

Earlier versions of material in some chapters appeared as articles in *Discover* and *Natural History* magazines. It is a pleasure to acknowledge my debt to many scientist colleagues for discussions and comments, to Roger Short and Nancy Wayne for their scrutiny of the whole manuscript, to Ellen Modecki for the illustrations, and to John Brockman for the invitation to write this book.

CHAPTER 1

· ·

THE ANIMAL WITH
THE WEIRDEST SEX LIFE

If your dog had your brain and could speak, and if you asked it what it thought of your sex life, you might be surprised by its response. It would be something like this:

> Those disgusting humans have sex any day of the month! Barbara proposes sex even when she knows perfectly well that she isn't fertile—like just after her period. John is eager for sex all the time, without caring whether his efforts could result in a baby or not. But if you want to hear something really gross—Barbara and John kept on having sex while she was pregnant! That's as bad as all the times when John's parents come for a visit, and I can hear them too having sex, although John's mother went through this thing they call menopause years ago. Now she can't have babies anymore, but she still wants sex, and John's father obliges her. What a waste of effort! Here's the weirdest thing of all: Barbara and John, and John's parents, close the bedroom door and have sex in private, instead of doing it in front of their friends like any self-respecting dog!

To understand where your dog is coming from, you need to free yourself from your human-based perspective on what constitutes normal sexual behavior. Increasingly today, we consider it narrow-minded and despicably prejudiced to denigrate those who do not conform to our own standards. Each such form of narrow-mindedness is associated with a despicable "ism"—for instance, racism, sexism, Eurocentrism, and phallocentrism. To that list of modern "ism" sins, defenders of animal rights are now adding the sin of species-ism. Our standards of sexual conduct are especially warped, species-ist, and human-centric because human sexuality is so abnormal by the standards of the world's thirty million other animal species. It's also abnormal by the standards of the world's millions of species of plants, fungi, and microbes, but I'll ignore that broader perspective because I haven't yet worked through my own zoo-centrism. This book confines itself to the insights that we can gain into our sexuality merely by broadening our perspective to encompass other animal species.

As a beginning, let's consider normal sexuality by the standards of the world's approximately 4,300 species of mammals, of which we humans are just one. Most mammals do not live as a nuclear family of a mated adult male and adult female, caring jointly for their offspring. Instead, in many mammal species both adult males and adult females are solitary, at least during the breeding season, and meet only to copulate. Hence, males do not provide paternal care; their sperm is their sole contribution to their offspring and to their temporary mate.

Even most social mammal species, such as lions, wolves, chimpanzees, and many hoofed mammals, are not paired off within the herd/pride/pack/band into male/female couples. Within such a herd/pride/et cetera, each adult male shows no signs of recognizing specific infants as his offspring by devoting himself to them at the expense of other infants in the herd. Indeed, it is only within the last few years that scientists studying lions, wolves, and chimpanzees have begun to figure out, with the help of

DNA testing, which male sired which infant. However, like all generalizations, these admit exceptions. Among the minority of adult male mammals that do offer their offspring paternal care are polygynous male zebras and gorillas with harems of females, male gibbons paired off with females as solitary couples, and saddleback tamarin monkeys, of which two adult males are kept as a harem by one polyandrous adult female.

Sex in social mammals is generally carried out in public, before the gazes of other members of the troop. For instance, a female Barbary macaque in estrus copulates with every adult male in her troop and makes no effort to conceal each copulation from other males. The best-documented exception to this pattern of public sex is in chimpanzee troops, where an adult male and estrous female may go off by themselves for a few days on what human observers term a "consortship." However, the same female chimpanzee that has private sex with a consort may also have public sex with other adult male chimpanzees within the same estrus cycle.

Adult females of most mammal species use various means of conspicuously advertising the brief phase of their reproductive cycle when they are ovulating and can be fertilized. The advertisement may be visual (for instance, the area around the vagina turning bright red), olfactory (releasing a distinctive smell), auditory (making noises), or behavioral (crouching in front of an adult male and displaying the vagina). Females solicit sex only during those fertile days, are sexually unattractive or less attractive to males on other days because they lack the arousing signals, and rebuff the advances of any male that is nevertheless interested on other days. Thus, sex is emphatically not just for fun and is rarely divorced from its function of fertilization. This generalization too admits exceptions: sex is flagrantly separated from reproduction in a few species, including bonobos (pygmy chimpanzees) and dolphins.

Finally, the existence of menopause as a regular phenomenon is not well established for most wild mammal
populations. By menopause is meant a complete cessation
of fertility within a time span that is much briefer than the
previous fertile career, and that is followed by an infertile
life span of significant length. Instead, wild mammals either are still fertile at the time of death or else exhibit gradually diminishing fertility with advancing age.

・・・・・

Now contrast what I have just said about normal mammalian sexuality with human sexuality. The following human attributes are among those that we take for granted as
normal:

1: Most men and women in most human societies end
up in a long-term pair relationship ("marriage") that other
members of the society recognize as a contract involving
mutual obligations. The couple has sex repeatedly, and
mainly or exclusively with each other.

2: In addition to being a sexual union, marriage is a
partnership for joint rearing of the resulting babies. In particular, human males as well as females commonly provide
parental care.

3: Despite forming a couple (or occasionally a harem), a
husband and wife (or wives) do not live (like gibbons) as a
solitary couple in an exclusive territory that they defend
against other couples, but instead they live embedded in a
society of other couples with whom they cooperate economically and share access to communal territory.

4: Marriage partners usually have sex in private, rather
than being indifferent to the presence of other humans.

5: Human ovulation is concealed rather than advertised. That is, women's brief period of fertility around the time of ovulation is difficult to detect for their potential sex partners as well as for most women themselves. A woman's sexual receptivity extends beyond the time of fertility to encompass most or all of the menstrual cycle. Hence, most human copulations occur at a time unsuitable for conception. That is, human sex is mostly for fun, not for insemination.

6: All women who live past the age of forty or fifty undergo menopause, a complete shutdown of fertility. Men in general do not undergo menopause: while individual men may develop fertility problems at any age, there is no age-clumping of infertility or universal shutdown.

Norms imply violation of norms: we call something a "norm" merely because it is more frequent than its opposite (the "violation of the norm"). That's as true for human sexual norms as for other norms. Readers of the last two pages will surely have been thinking of exceptions to the supposed generalizations that I have been describing, but they still stand as generalizations. For example, even in societies that recognize monogamy by law or custom there is much extramarital and premarital sex, and much sex that is not part of a long-term relationship. Humans do engage in one-night stands. On the other hand, most humans also engage in many-year or many-decade stands, whereas tigers and orangutans engage in nothing except one-night stands. The genetically based paternity tests developed over the last half-century have shown that the majority of American, British, and Italian babies are indeed sired by the husband (or steady boyfriend) of the baby's mother.

Readers may also bristle at hearing human societies described as monogamous; the term "harem," which zoologists apply to zebras and gorillas, is taken from the Arabic word for a human institution. Yes, many humans practice sequential monogamy. Yes, polygyny (long-term simultaneous unions between one man and multiple wives) is legal in some countries today, and polyandry (long-term simultaneous unions between one woman and multiple husbands) is legal in a few societies. In fact, polygyny was accepted in the great majority of traditional human societies before the rise of state institutions. However, even in officially polygynous societies most men have only one wife at a time, and only especially wealthy men can acquire and maintain a few wives simultaneously. The large harems that spring to mind at the mention of the word *polygamy,* such as those of recent Arabian and Indian royalty, are possible only in the state-level societies that arose very late in human evolution and that permitted a few men to concentrate great wealth. Hence the generalization stands: most adults in most human societies are at any given moment involved in a long-term pair bond that is often monogamous in practice as well as legally.

Still another cause for bristling may have been my description of human marriage as a partnership for the joint rearing of the resulting babies. Most children receive more parental care from their mothers than from their fathers. Unwed mothers form a significant proportion of the adult population in some modern societies, though it has been much harder for unwed mothers to rear children successfully in traditional societies. But the generalization again holds: most human children receive some parental care from their father, in the form of child care, teaching, protection, and provision of food, housing, and money.

All these features of human sexuality—long-term sexual partnerships, coparenting, proximity to the sexual partnerships of others, private sex, concealed ovulation, extended female receptivity, sex for fun, and female menopause—constitute what we humans assume is normal sexuality. It titillates, amuses, or disgusts us to read of the sexual habits of elephant seals, marsupial mice, or orangutans, whose lives are so different from ours. Their lives seem to us bizarre. But that proves to be a species-ist interpretation. By the standards of the world's 4,300 other species of mammals, and even by the standards of our own closest relatives, the great apes (the chimpanzee, bonobo, gorilla, and orangutan), we are the ones who are bizarre.

However, I am still being worse than zoo-centric. I am falling into the even narrower trap of mammalo-centrism. Do we become more normal when judged by the standards of nonmammalian animals? Other animals do exhibit a wider range of sexual and social systems than do mammals alone. Whereas the young of most mammal species receive maternal care but no paternal care, the reverse is true for some species of birds, frogs, and fish in which the father is the sole caretaker for his offspring. The male is a parasitic appendage fused to the female's body in some species of deep-sea fish; he is eaten by the female immediately after copulation in some species of spiders and insects. While humans and most other mammal species breed repeatedly, salmon, octopus, and many other animal species practice what is termed big-bang reproduction, or semelparity: a single reproductive effort, followed by preprogrammed death. The mating system of some species of birds, frogs, fish, and insects (as well as some bats and antelope) resembles a singles bar—at a traditional site, termed a "lek," many males maintain stations and compete for the attention of visiting females, each of which chooses a mate (often the same preferred male chosen by many other females), copulates with him, and then goes off to rear the resulting offspring without his assistance.

Among other animal species, it is possible to point out some whose sexuality resembles ours in particular respects. Most European and North American bird species form pair bonds that last for at least one breeding season (in some cases for life), and the father as well as the mother cares for the young. While most such bird species differ from us in that pairs occupy mutually exclusive territories, most species of sea birds resemble us further in that mated pairs breed colonially in close proximity to each other. However, all these bird species differ from us in that ovulation is advertised, female receptivity and the sex act are mostly confined to the fertile period around ovulation, sex is not recreational, and economic cooperation between pairs is slight or nonexistent. Bonobos (pygmy chimpanzees) resemble or approach us in many of these latter respects: female receptivity is extended through several weeks of the estrus cycle, sex is mainly recreational, and there is some economic cooperation between many members of the band. However, bonobos still lack our pair-bonded couples, our well-concealed ovulation, and our paternal recognition of and care for offspring. Most or all of these species differ from us in lacking a well-defined female menopause.

·····

Thus, even a non-mammalo-centric view reinforces our dog's interpretation: we are the ones who are bizarre. We marvel at what seems to us the weird behavior of peacocks and big-bang marsupial mice, but those species actually fall securely within the range of animal variation, and in fact we are the weirdest of them all. Species-ist zoologists theorize about why hammer-headed fruit bats evolved their lek mating system, yet the mating system that cries out for explanation is our own. Why did we evolve to be so different?

This question becomes even more acute when we compare ourselves with our closest relatives among the world's mammal species, the great apes (as distinguished from the gibbons or little apes). Closest of all are Africa's chimpanzee and bonobo, from which we differ in only about 1.6 percent of our nuclear genetic material (DNA). Nearly as close are the gorilla (2.3 percent genetic difference from us) and the orangutan of Southeast Asia (3.6 percent different). Our ancestors diverged "only" about seven million years ago from the ancestors of chimpanzees and bonobos, nine million years ago from the ancestors of gorillas, and fourteen million years ago from the ancestors of orangutans.

That sounds like an enormous amount of time in comparison to an individual human lifetime, but it's a mere eye-blink on the evolutionary time scale. Life has existed on Earth for more than three billion years, and hard-shelled, complex large animals exploded in diversity more than half a billion years ago. Within that relatively short period during which our ancestors and the ancestors of our great ape relatives have been evolving separately, we have diverged in only a few significant respects and to a modest degree, even though some of those modest differences— especially our upright posture and larger brains—have had enormous consequences for our behavioral differences.

Along with posture and brain size, sexuality completes the trinity of the decisive respects in which the ancestors of humans and great apes diverged. Orangutans are often solitary, males and females associate just to copulate, and males provide no paternal care; a gorilla male gathers a harem of a few females, with each of which he has sex at intervals of several years (after the female weans her most recent offspring and resumes menstrual cycling and before she becomes pregnant again); and chimpanzees and bonobos live in troops with no lasting male-female pair bonds or specific father-offspring bonds. It is clear how our large brain and upright posture played a decisive role in what is termed our humanity—in the fact that we now use language, read books, watch TV, buy or grow most of our food, occupy all continents and oceans, keep members of our own and other species in cages, and are exterminating most other animal and plant species, while the great apes still speechlessly gather wild fruit in the jungle, occupy small ranges in the Old World tropics, cage no animal, and threaten the existence of no other species. What role did our weird sexuality play in our achieving these hallmarks of humanity?

Could our sexual distinctiveness be related to our other distinctions from the great apes? In addition to (and probably ultimately as a product of) our upright posture and large brains, those distinctions include our relative hairlessness, dependence on tools, command of fire, and development of language, art, and writing. If any of these distinctions predisposed us toward evolving our sexual distinctions, the links are certainly unclear. For example, it is not obvious why our loss of body hair should have made recreational sex more appealing, nor why our command of fire should have favored menopause. Instead, I shall argue the reverse: recreational sex and menopause were as important for our development of fire, language, art, and writing as were our upright posture and large brains.

·····

The key to understanding human sexuality is to recognize that it is a problem in evolutionary biology. When Darwin recognized the phenomenon of biological evolution in his great book *On the Origin of Species,* most of his evidence was drawn from anatomy. He inferred that most plant and animal structures evolve—that is, they tend to change from generation to generation. He also inferred that the major force behind evolutionary change is natural selection. By that term, Darwin meant that plants and animals vary in their anatomical adaptations, that certain adaptations enable individuals bearing them to survive and reproduce more successfully than other individuals, and that those particular adaptations therefore increase in frequency in a population from generation to generation. Later biologists showed that Darwin's reasoning about anatomy also applies to physiology and biochemistry: an animal's or plant's physiological and biochemical characteristics also adapt it to certain lifestyles and evolve in response to environmental conditions.

More recently, evolutionary biologists have shown that animal social systems also evolve and adapt. Even among closely related animal species, some are solitary, others live in small groups, and still others live in large groups. But social behavior has consequences for survival and reproduction. Depending, for example, on whether a species' food supply is clumped or spread out, and on whether a species faces high risk of attack by predators, either solitary living or group living may be better for promoting survival and reproduction.

Similar considerations apply to sexuality. Some sexual characteristics may be more advantageous for survival and reproduction than others, depending on each species' food supply, exposure to predators, and other biological characteristics. At this point I shall mention just one example, a behavior that at first seems diametrically opposed to evolutionary logic: sexual cannibalism. The male of some species of spiders and mantises is routinely eaten by his mate just after or even while he is copulating with her. This cannibalism clearly involves the male's consent, because the male of these species approaches the female, makes no attempt to escape, and may even bend his head and thorax toward the female's mouth so that she may munch her way through most of his body while his abdomen remains to complete the job of injecting sperm into her.

If one thinks of natural selection as the maximization of survival, such cannibalistic suicide makes no sense. Actually, natural selection maximizes the transmission of genes, and survival is in most cases just one strategy that provides repeated opportunities to transmit genes. Suppose that opportunities to transmit genes arise unpredictably and infrequently, and that the number of offspring produced by such opportunities increases with the female's nutritional condition. That's the case for some species of spiders and mantises living at low population densities. A male is lucky to encounter a female at all, and such luck is unlikely to strike twice. The male's best strategy is to produce as many offspring bearing his genes as possible out of his lucky find. The larger a female's nutritional reserves, the more calories and protein she has available to transform into eggs. If the male departed after mating, he would probably not find another female and his continued survival would thus be useless. Instead, by encouraging the female to eat him, he enables her to produce more eggs bearing his genes. In addition, a female spider whose mouth is distracted by munching a male's body allows copulation with the male's genitalia to proceed for a longer time, resulting in more sperm transferred and more eggs fertilized. The male spider's evolutionary logic is impeccable and seems bizarre to us only because other aspects of human biology make sexual cannibalism disadvantageous. Most men have more than one lifetime opportunity to copulate; even well-nourished women usually give birth to only a single baby at a time, or at most twins; and a woman could not consume enough of a man's body at one sitting to improve significantly the nutritional basis for her pregnancy.

This example illustrates the dependence of evolved sexual strategies on both ecological parameters and the parameters of a species' biology, both of which vary among species. Sexual cannibalism in spiders and mantises is favored by the ecological variables of low population densities and low encounter rates, and by the biological variables of a female's capacity to digest relatively large meals and to increase her egg output considerably when well nourished. Ecological parameters can change overnight if an individual colonizes a new type of habitat, but the colonizing individual carries with it a baggage of inherited biological attributes that can change only slowly, through natural selection. Hence it is not enough to consider a species' habitat and lifestyle, design on paper a set of sexual characteristics that would be well matched to that habitat and lifestyle, and then be surprised that those supposedly optimal sexual characteristics do not evolve. Instead, sexual evolution is severely constrained by inherited commitments and prior evolutionary history.

For example, in most fish species a female lays eggs and a male fertilizes those eggs outside the female's body, but in all placental mammal species and marsupials a female gives birth to live young rather than to eggs, and all mammal species practice internal fertilization (male sperm injected into the female's body). Live birth and internal fertilization involve so many biological adaptations and so many genes that all placental mammals and marsupials have been firmly committed to those attributes for tens of millions of years. As we shall see, these inherited commitments help explain why there is no mammal species in which parental care is provided solely by the male, even in habitats where mammals live alongside fish and frog species whose males are the sole providers of parental care.

We can thus redefine the problem posed by our strange sexuality. Within the last seven million years, our sexual anatomy diverged somewhat, our sexual physiology further, and our sexual behavior even more, from those of our closest relatives, the chimpanzees. Those divergences must reflect a divergence between humans and chimpanzees in environment and lifestyle. But those divergences were also limited by inherited constraints. What were the lifestyle changes and inherited constraints that molded the evolution of our weird sexuality?

··

CHAPTER 2

THE BATTLE OF THE SEXES

In the preceding chapter we saw that our effort to understand human sexuality must begin by our distancing ourselves from our warped human perspective. We're exceptional animals in that our fathers and mothers often remain together after copulating and are both involved in rearing the resulting child. No one could claim that men's and women's parental contributions are equal: they tend to be grossly unequal in most marriages and societies. But most fathers make some contribution to their children, even if it's just food or defense or land rights. We take such contributions so much for granted that they're written into law: divorced fathers owe child support, and even an unwed mother can sue a man for child support if genetic testing proves that he is her child's father.

But that's our warped human perspective. Alas for sexual equality, we're aberrations in the animal world, and especially among mammals. If orangutans, giraffes, and most other mammal species could express their opinion, they would declare our child support laws absurd. Most male mammals have no involvement with either their offspring or their offspring's mother after inseminating her; they are too busy seeking other females to inseminate. Male animals in general, not just male mammals, provide much less parental care (if any) than do females.

Yet there are quite a few exceptions to this chauvinist pattern. In some bird species, such as phalaropes and Spotted Sandpipers, it's the male that does the work of incubating the eggs and rearing the chicks, while the female goes in search of another male to inseminate her again and to rear her next clutch. Males of some fish species (like seahorses and sticklebacks) and some amphibian males (like midwife toads) care for the eggs in a nest or in their mouth, pouch, or back. How can we explain simultaneously this general pattern of female parental care and also its numerous exceptions?

The answer comes from the realization that genes for behavior, as well as for malaria resistance and teeth, are subject to natural selection. A behavior pattern that helps individuals of one animal species pass on their genes won't necessarily be helpful in another species. In particular, a male and female that have just copulated to produce a fertilized egg face a "choice" of subsequent behaviors. Should that male and female both leave the egg to fend for itself and set to work on producing another fertilized egg, copulating either with the same partner or with a different partner? On the one hand, a time-out from sex for the purpose of parental care might improve the chances of the first egg surviving. If so, that choice leads to further choices: both the mother and the father could choose to provide the parental care, or just the mother could choose to do so, or just the father could. On the other hand, if the egg has a one-in-ten chance of surviving even with no parental care, and if the time you'd devote to tending it would alternatively let you produce 1,000 more fertilized eggs, you'd be best off leaving that first egg to fend for itself and going on to produce more fertilized eggs.

I've referred to these alternatives as "choices." That word may seem to suggest that animals operate like human decision-makers, consciously evaluating alternatives and finally choosing the particular alternative that seems most likely to advance the animal's self-interest. Of course, that's not what happens. Many of the so-called choices actually are programmed into an animal's anatomy and physiology. For example, female kangaroos have "chosen" to have a pouch that can accommodate their young, but male kangaroos have not. Most or all of the remaining choices are ones that would be anatomically possible for either sex, but animals have programmed instincts that lead them to provide (or not to provide) parental care, and this instinctive "choice" of behavior can differ between sexes of the same species. For example, among parent birds, both male and female albatrosses, male but not female ostriches, females but not males of most hummingbird species, and no brush turkeys of either sex are instinctively programmed to bring food to their chicks, although both sexes of all of these species are physically and anatomically perfectly capable of doing so.

The anatomy, physiology, and instincts underlying parental care are all programmed genetically by natural selection. Collectively, they constitute part of what biologists term a reproductive strategy. That is, genetic mutations or recombinations in a parent bird could strengthen or weaken the instinct to bring food to the chicks and could do so differently in the two sexes of the same species. Those instincts are likely to have a big effect on the number of chicks that survive to carry on the parent's genes. It's obvious that a chick to which a parent brings food is more likely to survive, but we shall also see that a parent that *forgoes* bringing food to its chicks thereby gains other increased chances to pass on its genes. Hence the net effect of a gene that causes a parent bird instinctively to bring food to its chicks could be either to increase or to decrease the number of chicks carrying on the parent's genes, depending on ecological and biological factors that we shall discuss.

Genes that specify the particular anatomical structures

or instincts most likely to ensure the survival of offspring bearing the genes will tend to increase in frequency. This statement can be rephrased: anatomical structures and instincts that promote survival and reproductive success tend to become established (genetically programmed) by natural selection. But the need to make wordy statements such as these arises very often in any discussion of evolutionary biology. Hence biologists routinely resort to anthropomorphic language to condense such statements—for example, they say that an animal "chooses" to do something or pursues a certain strategy. This shorthand vocabulary should not be misconstrued as implying that animals make conscious calculations.

· · · · ·

For a long time, evolutionary biologists thought of natural selection as somehow promoting "the good of the species." In fact, natural selection operates initially on individual animals and plants. Natural selection is not just a struggle between species (entire populations), nor is it just a struggle between individuals of different species, nor just between conspecific individuals of the same age and sex. Natural selection can also be a struggle between parents and their offspring or a struggle between mates, because the self-interests of parents and their offspring, or of father and mother, may not coincide. What makes individuals of one age and sex successful at transmitting their genes may not increase the success of other classes of individuals.

In particular, while natural selection favors both males and females that leave many offspring, the best strategy for doing so may be different for fathers and mothers. That generates a built-in conflict between the parents, a conclusion that all too many humans don't need scientists to reveal to them. We make jokes about the battle of the sexes, but the battle is neither a joke nor an aberrant accident of

how individual father or mothers behave on particular oc-
casions. It is indeed perfectly true that behavior that is in a
male's genetic interests may not necessarily be in the interests of his female co-parent, and vice versa. That cruel fact
is one of the fundamental causes of human misery.

Consider again the case of the male and female that
have just copulated to produce a fertilized egg and now
face the "choice" of what to do next. If the egg has some
chance of surviving unassisted, and if both the mother and
the father could produce many more fertilized eggs in the
time that they would devote to tending that first fertilized
egg, then the interests of the mother and father coincide in
deserting the egg. But now suppose that the newly fertilized, laid, or hatched egg or newborn offspring has ab-
solutely zero chance of surviving unless it is cared for by
one parent. Then there is indeed a conflict of interest.
Should one parent succeed in foisting the obligation of
parental care onto the other parent and then going off in
search of a new sex partner, then the foister will have advanced her or his genetic interests at the expense of the
abandoned parent. The foister will really promote his or
her selfish evolutionary goals by deserting his or her mate
and offspring.

In such cases when care by one parent is essential for
offspring survival, child-rearing can be thought of as a
cold-blooded race between mother and father to be the first
to desert the other and their mutual offspring and to get on
with the business of producing more babies. Whether it actually pays you to desert depends on whether you can
count on your old mate to finish rearing the kids, and
whether you are then likely to find a receptive new mate.
It's as if, at the moment of fertilization, the mother and father play a game of chicken, stare at each other, and simultaneously say, "I am going to walk off and find a new
partner, and you can care for this embryo if you want to,
but even if you don't, *I won't!*" If both partners call each

other's bluff in that race to desert their embryo, then the embryo dies and both parents lose the game of chicken. Which parent is more likely to back down?

The answer depends on such considerations as which parent has more invested in the fertilized egg, and which parent has better alternative prospects. As I said before, neither parent makes a conscious calculation; the actions of each parent are instead programmed genetically by natural selection into the anatomy and instincts of their sex. In many animal species the female backs down and becomes sole parent while the male deserts, but in other species the male assumes responsibility and the female deserts, and in still other species both parents assume shared responsibility. Those varying outcomes depend on three interrelated sets of factors whose differences between the sexes vary among species: investment in the already fertilized embryo or egg; alternative opportunities that would be foreclosed by further care of the already fertilized embryo or egg; and confidence in the paternity or maternity of the embryo or egg.

·····

All of us know from experience that we are much more reluctant to walk away from an ongoing enterprise in which we have invested a lot than from one in which we have invested only a little. That's true of our investments in human relationships, in business projects, or in the stock market. It's true regardless of whether our investment is in the form of money, time, or effort. We lightly end a relationship that turns bad on the first date, and we stop trying to construct from parts a cheap toy when we hit a snag within a few minutes. But we agonize over ending a twenty-five-year marriage or an expensive house remodeling.

The same principle applies to parental investment in potential offspring. Even at the moment when an egg is fertilized by a sperm, the resulting fertilized embryo generally represents a greater investment for the female than for the male, because in most animal species the egg is much larger than the sperm. While both eggs and sperm contain chromosomes, the egg in addition must contain enough nutrients and metabolic machinery to support the embryo's further development for some time, at least until the embryo can start feeding itself. Sperm, in contrast, need contain only a flagellar motor and sufficient energy to drive that motor and support swimming for at most a few days. As a result, a mature human egg has roughly one million times the mass of the sperm that fertilizes it; the corresponding factor for kiwis is one million billion. Hence a fertilized embryo, viewed simply as an early-stage construction project, represents an utterly trivial investment of its father's body mass compared to its mother's. But that doesn't mean that the female has automatically lost the game of chicken before the moment of conception. Along with the one sperm that fertilized the egg, the male may have produced several hundred million other sperm in the ejaculate, so that his total investment may be not dissimilar to the female's.

The act of fertilizing an egg is described as either internal or external, depending on whether it takes place inside or outside the female's body. External fertilization characterizes most species of fish and amphibia. For example, in most fish species a female and a nearby male simultaneously discharge their eggs and sperm into the water, where fertilization occurs. With external fertilization, the female's obligate investment ends at the moment she extrudes the eggs. The embryos may then be left to float away and fend for themselves without parental care, or they may receive care from one parent, depending on the species.

More familiar to humans is internal fertilization, the male's injection of sperm (for example, via an intromittive penis) into the female's body. What happens next in most species is that the female does not immediately extrude the embryos but retains them in her body for a period of development until they are closer to the stage when they can survive by themselves. The offspring may eventually be packaged for release within a protective eggshell, together with an energy supply in the form of yolk—as in all birds, many reptiles, and monotreme mammals (the platypus and echidnas of Australia and New Guinea). Alternatively, the embryo may continue to grow within the mother until the embryo is "born" without an eggshell instead of being "laid" as an egg. That alternative, termed vivipary (Latin for "live birth"), characterizes us and all other mammals except monotremes, plus some fish, reptiles, and amphibia. Vivipary requires specialized internal structures—of which the mammalian placenta is the most complex—for the transfer of nutrients from the mother to her developing embryo and the transfer of wastes from embryo to mother.

Internal fertilization thus obligates the mother to further investment in the embryo beyond the investment that she has already made in producing the egg until it is fertilized. Either she uses calcium and nutrients from her own body to make an eggshell and yolk, or else she uses her nutrients to make the embryo's body itself. Besides that investment of nutrients, the mother is also obligated to invest the time required for pregnancy. The result is that the investment of an internally fertilized mother at the time of hatching or birth, relative to the father's, is likely to be much greater that that of an externally fertilized mother at the time of unfertilized egg extrusion. For instance, by the end of a nine-month pregnancy a human mother's expenditure of time and energy is colossal in comparison with her husband's or boyfriend's pathetically slight investment during the few minutes it took him to copulate and extrude his one milliliter of sperm.

As a result of that unequal investment of mothers and fathers in internally fertilized embryos, it becomes harder for the mother to bluff her way out of post-hatching or post-birth parental care, if any is required. That care takes many forms: for instance, lactation by female mammals, guarding the eggs by female alligators, and brooding the eggs by female pythons. Nevertheless, as we shall see, there are other circumstances that may induce the father to stop bluffing and to start assuming shared or even sole responsibility for his offspring.

· · · · ·

I mentioned that three related sets of factors influence the "choice" of parent to be caretaker, and that relative size of investment in the young is only one of those factors. A second factor is foreclosed opportunity. Picture yourself as an animal parent contemplating your newborn offspring and coldly calculating your genetic self-interest as you debate what you should now do with your time. That offspring bears your genes, and its chance of surviving to perpetuate your genes would undoubtedly be improved if you hung around to protect and feed it. If there is nothing else you could do with your time to perpetuate your genes, your interests would be best served by caring for that offspring and not trying to bluff your mate into being sole parent. On the other hand, if you can think of ways to spread your genes to many more offspring in the same time, you should certainly do so and desert your current mate and offspring.

Now consider a mother and father animal both doing that calculation the moment after they have mated to produce some fertilized embryos. If fertilization is external, neither mother nor father is automatically committed to anything further, and both are theoretically free to seek another partner with whom to produce more fertilized embryos. Yes, their just-fertilized embryos may need some care, but mother and father are equally able to try to bluff the other into providing that care. But if fertilization is internal, the female is now pregnant and committed to nourishing the fertilized embryos until birth or laying. If she is a mammal, she is committed for even longer, through the period of lactation. During that period it does her no genetic good to copulate with another male, because she cannot thereby produce more babies. That is, she loses nothing by devoting herself to child care.

But the male who has just discharged his sperm sample into one female is available a moment later to discharge another sperm sample into another female, and thereby potentially to pass his genes to more offspring. A man, for example, produces about two hundred million sperm in one ejaculate—or at least a few tens of millions, even if reports of a decline in human sperm count in recent decades are correct. By ejaculating once every 28 days during his recent partner's 280-day pregnancy—a frequency of ejaculation easily within the reach of most men—he would broadcast enough sperm to fertilize every one of the world's approximately two billion reproductively mature women, if he could only succeed in arranging for each of them to receive one of his sperm. That's the evolutionary logic that induces so many men to desert a woman immediately after impregnating her and to move on to the next woman. A man who devotes himself to child care potentially forecloses many alternative opportunities. Similar logic applies to males and females of most other internally fertilized animals. Those alternative opportunities available to males contribute to the predominant pattern of females providing child care in the animal world.

The remaining factor is confidence of parenthood. If you are going to invest time, effort, and nutrients in raising a fertilized egg or embryo, you'd better make damn sure first that it's your own offspring. If it turns out to be somebody else's offspring, you've lost the evolutionary race. You'll have knocked yourself out in order to pass on a rival's genes.

For women and other female animals practicing internal fertilization, doubt about maternity never arises. Into the mother's body, containing her eggs, goes sperm. Out of her body sometime later comes a baby. There's no way that the baby could have been switched with some other mother's baby inside of her. It's a safe evolutionary bet for the mother to care for that baby.

But males of mammals and other internally fertilized animals have no corresponding confidence in their paternity. Yes, the male knows that his sperm went into a female's body. Sometime later, out of that female's body, comes a baby. How does the male know whether the female copulated with other males while he wasn't looking? How does he know whether his sperm or some other male's sperm was the one that fertilized the egg? In the face of this inevitable uncertainty, the evolutionary conclusion reached by most male mammals is to walk off the job immediately after copulation, seek more females to impregnate, and leave those females to rear their offspring— hoping that one or more of the females with which he copulated will actually have been impregnated by him and will succeed in rearing his offspring unassisted. Male parental care would be a bad evolutionary gamble.

. . . . .

Yet we know, from our own experience, that some species constitute exceptions to that general pattern of male postcopulatory desertion. The exceptions are of three types. One type is those species whose eggs are fertilized externally. The female ejects her not yet fertilized eggs; the male, hovering nearby or already grasping the female, spreads his sperm on the eggs; he immediately scoops up the eggs, before any other males have a chance to cloud the picture with their sperm; and he proceeds to care for the eggs, completely confident in his paternity. This is the evolutionary logic that programs some male fish and frogs to play the role of sole parent after fertilization. For example, the male midwife toad guards the eggs by wrapping them around his hind legs; the male glass frog stands watch over eggs in vegetation over a stream into which the hatched tadpoles can drop; and the male stickleback builds a nest in which to protect the eggs against predators.

A second type of exception to the predominant pattern of male post-copulatory desertion involves a remarkable phenomenon with a long name: sex-role-reversal polyandry. As the name implies, this behavior is the opposite of the common polygynous breeding systems in which big males compete fiercely with each other to acquire a harem of females. Instead, big females compete fiercely to acquire a harem of smaller males, for each of which in turn the female lays a clutch of eggs, and each of which proceeds to do most or all of the work of incubating the eggs and rearing the young. The best known of these female sultans are the shore birds called jacanas (alias lily-trotters), Spotted Sandpipers, and Wilson's Phalaropes. For instance, flocks of up to ten female phalaropes may pursue a male for miles. The victorious female then stands guard over her prize to ensure that only she gets to have sex with him, and that he becomes one of the males rearing her chicks.

Clearly, sex-role-reversal polyandry represents for the successful female the fulfillment of an evolutionary dream. She wins the battle of the sexes by passing on her genes to far more clutches of young than she could rear, alone or with one male's help. She can utilize nearly her full egg-laying potential, limited only by her ability to defeat other females in the quest for males willing to take over parental care. But how did this strategy evolve? Why did males of some shorebird species end up seemingly defeated in the battle of the sexes, as polyandrous co-"husbands," when males of almost all other bird species avoided that fate or even reversed it to become polygynists?

The explanation depends on shorebirds' unusual reproductive biology. They lay only four eggs at a time, and the young are precocial, meaning that they hatch already covered with down, with their eyes open, and able to run and find food for themselves. The parent doesn't have to feed the chicks but only has to protect them and keep them warm. That's something a single parent can handle, whereas it takes two parents to feed the young of most other bird species.

But a chick that can run around as soon as it hatches has undergone more development inside the egg than the usual helpless chick. That requires an exceptionally large egg. (Take a look sometime at a pigeon's typically small eggs, which produce the usual helpless chicks, to understand why egg farmers prefer to rear chickens with big eggs and precocial chicks.) In Spotted Sandpipers, each egg weighs fully one-fifth as much as its mother; the whole four-egg clutch weighs an astonishing 80 percent of her weight. Although even monogamous shorebird females have evolved to be slightly larger than their mates, the effort of producing those huge eggs is still exhausting. That maternal effort gives the male both a short-term and a long-term advantage if he takes over the not too onerous responsibility of rearing the precocial chicks alone, thereby leaving his mate free to fatten herself up again.

His short-term advantage is that his mate thereby becomes capable of producing another clutch of eggs for him quickly, in case the first clutch is destroyed by a predator. That's a big advantage, because shorebirds nest on the ground and suffer horrendous losses of eggs and chicks. For example, in 1975 a single mink destroyed every nest in a population of Spotted Sandpipers that the ornithologist Lewis Oring was studying in Minnesota. A study of jacanas in Panama found that forty-four out of fifty-two nests failed.

Sparing his mate may also bring the male a long-term advantage. If she does not become exhausted in one breeding season, she is more likely to survive to the next season, when he can mate with her again. Like human couples, experienced bird couples that have worked out a harmonious relationship are more successful at raising young than are bird newlyweds.

But generosity in anticipation of later repayment carries a risk, for male shorebirds as for humans. Once the male assumes sole parental responsibility, the road is clear for his mate to use her free time in whatever way she chooses. Perhaps she'll choose to reciprocate and remain available to her mate, on the chance that her first clutch might be destroyed and he would require a replacement clutch. But she might also choose to pursue her own interests, seeking out some other male available immediately to receive her second clutch. If her first clutch survives and continues to occupy her former mate, her polyandrous strategy has thereby doubled her genetic output.

Naturally, other females will have the same idea, and all of them will find themselves in competition for a dwindling supply of males. As the breeding season progresses, most males become tied up with their first clutch and unable to accept further parental responsibilities. Although the numbers of adult males and females may be equal, the ratio of sexually *available* females to males rises as high as seven-to-one among breeding Spotted Sandpipers and Wilson's Phalaropes. Those cruel numbers are what drive sex-role reversal even further toward an extreme. Though females already had to be slightly larger than males in order to produce large eggs, they have evolved to become still larger in order to win the fights with other females. The female reduces her own parental care contribution further and woos the male rather than vice versa.

Thus, the distinctive features of shorebird biology—especially their precocial young, clutches of few but large eggs, ground-nesting habits, and severe losses from predation—predispose them to male uniparental care and female emancipation or desertion. Granted, females of most shorebird species can't exploit those opportunities for polyandry. That's true, for instance, of most sandpipers of the high Arctic, where the very short breeding season leaves no time for a second clutch to be reared. Only among a minority of species, such as the tropical jacanas and southerly populations of Spotted Sandpipers, is polyandry frequent or routine. Though seemingly remote from human sexuality, shorebird sexuality is instructive because it illustrates the main message of this book: a species' sexuality is molded by other aspects of the species' biology. It's easier for us to acknowledge this conclusion about shorebirds, to which we don't apply moral standards, than about ourselves.

•••••

The remaining type of exception to the predominant pattern of male desertion occurs in species in which, like us, fertilization is internal but it's hard or impossible for a single parent to rear the young unassisted. A second parent may be required to gather food for the coparent or the young, tend the young while the coparent is off gathering food, defend a territory, or teach the young. In such species the female alone would not be able to feed and defend the young without the male's help. Deserting a fertilized mate to pursue other females would bring no evolutionary gain to a male if his offspring thereby died of starvation. Thus, self-interest may force the male to remain with his fertilized spouse, and vice versa.

That's the case with most of our familiar North American and European birds: males and females are monogamous, and they share in caring for the young. It's also approximately true for humans, as we know so well. Human single-parenthood is difficult enough, even in these days of supermarket shopping and babysitters for hire. In ancient hunter-gatherer days, a child orphaned by either its mother's or its father's death faced reduced chances of survival. The father as well as the mother desirous of passing on genes finds it a matter of self-interest to care for the child. Hence most men have provided food, protection, and housing for their spouse and kids. The result is our human social system of nominally monogamous married couples, or occasionally of harems of women committed to one affluent man. Essentially the same considerations apply to gorillas, gibbons, and the other minority mammals practicing male parental care.

Yet that familiar arrangement of coparenthood does not end the battle of the sexes. It does not necessarily dissolve the tension between the mother's and father's interests, arising from their unequal investments before birth. Even among those mammal and bird species that provide paternal care, males try to see how little care they can get away with and still have the offspring survive owing mainly to the mother's efforts. Males also try to impregnate other males' mates, leaving the unfortunate cuckolded male to care unknowingly for the cuckolder's offspring. Males become justifiably paranoid about their mates' behavior.

An intensively studied and fairly typical example of those built-in tensions of coparenthood is the European bird species known as the Pied Flycatcher. Most flycatcher males are nominally monogamous, but many try to be polygynous, and quite a few succeed. Again, it is instructive to devote a few pages of this book on human sexuality to another example involving birds, because (as we'll see) the behavior of some birds is strikingly like that of humans but does not arouse the same moral indignation in us.

Here is how polygyny works for Pied Flycatchers. In the spring a male finds a good nest hole, stakes out his territory around it, woos a female, and copulates with her. When this female (termed his primary female) lays her first egg, the male feels confident that he has fertilized her, that she'll be busy incubating his eggs, and that she won't be interested in other males and is temporarily sterile anyway. Hence the male finds another nest hole nearby, courts another female (termed his secondary female), and copulates with her.

When that secondary female begins laying, the male feels confident that he has fertilized her as well. Around that same time, the eggs of his primary female are starting to hatch. The male returns to her, devotes most of his energy to feeding her chicks and devotes less or no energy to feeding the chicks of his secondary female. Numbers tell the cruel story: the male averages fourteen deliveries of food per hour to the primary female's nest but only seven deliveries of food per hour to the secondary female's nest. If enough nest holes are available, most mated males try to acquire a secondary female, and up to 39 percent succeed.

Obviously, this system produces both winners and losers. Since the numbers of male and female flycatchers are roughly equal, and since each female has one mate, for every bigamous male there must be one unfortunate male with no mate. The big winners are the polygynous males, who sire on the average 8.1 flycatcher chicks each year (adding up the contributions of both mates), compared to only 5.5 chicks sired by monogamous males. Polygynous males tend to be older and bigger than unmated males, and they succeed in staking out the best territories and best nest holes in the best habitats. As a result, their chicks end up 10 percent heavier than the chicks of other males, and those big chicks have a better chance of surviving than do smaller chicks.

The biggest losers are the unfortunate unmated males, who fail to acquire any mates and sire no offspring at all (at least in theory—more on that later). The other losers are the secondary females, who have to work much harder than primary females to feed their young. The former end up making twenty food deliveries per hour to the nest, compared with only thirteen for the latter. Since the secondary females thus exhaust themselves, they may die earlier. Despite her herculean efforts, one hardworking secondary female can't bring as much food to the nest as a relaxed primary female and a male working together. Hence some chicks starve, and the secondary females end up with fewer surviving chicks than do primary females (on the average, 3.4 versus 5.4 chicks). In addition, the surviving chicks of secondary females are smaller than the chicks of primary females, and hence are less likely to survive the rigors of winter and migration.

Given these cruel statistics, why should any female accept the fate of being the "other woman"? Biologists used to speculate that secondary females choose their fate, reasoning that the neglected second spouse of a good male is better off than the sole spouse of a lousy male with a poor territory. (Rich married men have been known to make similar pitches to prospective mistresses.) It turns out, though, that the secondary females do not accept their fate knowingly but are tricked into it.

The key to this deception is the care that polygynous males take to set up their second household a couple of hundred yards from their first household, with many other males' territories intervening. It's striking that polygynous males don't court a second spouse at any of dozens of potential nest holes near the first nest, even though they would thereby reduce their commuting time between nests, have more time available to feed their young, and reduce their risk of being cuckolded while en route. The conclusion seems inescapable that polygynous males accept the disadvantage of a remote second household in order to deceive the prospective secondary mate and conceal from her the existence of the first household. Life's exigencies make a female Pied Flycatcher especially vulnerable to being deceived. If she discovers after egg-laying that her mate is polygynous, it's too late for her to do anything about it. She's better off staying with those eggs than deserting them, seeking a new mate from the males now available (most of them are would-be bigamists anyway), and hoping the new mate will prove to be any better than the former one.

The remaining strategy of male Pied Flycatchers has been dressed up by male biologists in the morally neutral-sounding term "mixed reproductive strategy" (abbreviated MRS). What this means is that mated male Pied Flycatchers don't just have a mate: they also sneak around trying to inseminate the mates of other males. If they find a female whose mate is temporarily absent, they try to copulate with her and often succeed. Either they approach her singing loudly or they sneak up to her quietly; the latter method succeeds more often.

The scale of this activity staggers our human imagination. In act 1 of Mozart's opera *Don Giovanni,* the Don's servant, Leporello, boasts to Donna Elvira that Don Giovanni has seduced 1,003 women in Spain alone. That sounds impressive until you realize how long-lived we humans are. If Don Giovanni's conquests took place over thirty years, he seduced only one Spanish woman every eleven days. In contrast, if a male Pied Flycatcher temporarily leaves his mate (for instance, to find food), then on the average another male enters his territory in ten minutes and copulates with his mate in thirty-four minutes. Twenty-nine percent of all observed copulations prove to be EPCs (extra-pair copulations), and an estimated 24 percent of all nestlings are "illegitimate." The intruder-seducer usually proves to be the boy next door (a male from an adjoining territory).

The big loser is the cuckolded male, for whom EPCs and MRSs are an evolutionary disaster. He squanders a whole breeding season out of his short life by feeding chicks that do not pass on his genes. Although the male perpetrator of an EPC might seem to be the big winner, a little reflection makes it clear that working out the male's balance sheet is tricky. While you are off philandering, other males have the chance to philander with your mate. EPC attempts rarely succeed if a female is within ten yards of her mate, but the chances of success rise steeply if her mate is more distant than ten yards. That makes MRSs especially risky for polygynous males, who spend much time in their other territory or commuting between their two territories. The polygynous males try to pull off EPCs themselves and on the average make one attempt every twenty-five minutes, but once every eleven minutes some other male is sneaking into their own territory to try for an EPC. In half of all EPC attempts, the cuckolded male flycatcher is off in pursuit of another female flycatcher at the very moment when his own mate is under siege.

These statistics would seem to make MRSs a strategy of dubious value to male Pied Flycatchers, but they are clever enough to minimize their risks. Until they have fertilized their own mate, they stay within two or three yards of her and guard her diligently. Only when she has been inseminated do they go off philandering.

· · · · ·

Now that we have surveyed the varying outcomes of the battle of the sexes in animals, let's see how humans fit into this broader picture. While human sexuality is unique in other respects, it is quite ordinary when it comes to the battle of the sexes. Human sexuality resembles that of many other animal species whose offspring are internally fertilized and require biparental care. It thereby differs from that of most species whose young are externally fertilized and given only uniparental care or even no care at all.

In humans, as in all other mammalian and bird species except brush turkeys, an egg that has just been fertilized is incapable of independent survival. In fact, the length of time until the offspring can forage and care for itself is at least as long for humans as for any other animal species, and far longer than for the vast majority of animal species. Hence parental care is indispensable. The only question is, which parent will provide that care or will both parents provide it?

For animals, we saw that the answer to that question depends on the relative size of the mother's and father's obligate investment in the embryo, their other opportunities foreclosed by their choice to provide parental care, and their confidence in their paternity or maternity. Looking at the first of those factors, the human mother has a greater obligate investment than the human father. Already at the time of fertilization a human egg is much larger than a human sperm, though that discrepancy disappears or is reversed if the egg is compared to an entire ejaculate of sperm. After fertilization the human mother is committed to up to nine months of time and energy expenditure, followed by a period of lactation that lasted about four years under the conditions of the hunter-gatherer lifestyle that characterized all human societies until the rise of agriculture about ten thousand years ago. As I recall well myself from watching how fast the food disappeared from our refrigerator when my wife was nursing our sons, human lactation is energetically very expensive. The daily energy budget of a nursing mother exceeds that of most men with even a moderately active lifestyle and is topped among women only by marathon runners in training. Hence there is no way that a just-fertilized woman can rise from the conjugal bed, look her spouse or lover in the eye, and tell him, "You'll have to take care of this embryo if you want it to survive, because I won't!" Her consort would recognize this for an empty bluff.

The second factor affecting the relative interest of men and women in child care is their difference in other opportunities thereby foreclosed. Because of the woman's time commitment to pregnancy and (under hunter-gatherer conditions) lactation, there is nothing she can do during that time that would permit her to produce another offspring. The traditional nursing pattern was to nurse many times each hour, and the resulting release of hormones tended to cause lactational amenorrhea (cessation of menstrual cycles) for up to several years. Hence hunter-gatherer mothers had children at intervals of several years. In modern society a woman can conceive again within a few months of delivery, either by forgoing breast-feeding in favor of bottle-feeding or by nursing the infant only every few hours (as modern women tend to do for convenience). Under those conditions the woman soon resumes menstrual cycles. Nevertheless, even modern women who eschew breast-feeding and contraception rarely give birth at intervals of less than a year, and few women give birth to more than a dozen children over the course of their lives. The record lifetime number of offspring for a woman is a mere sixty-nine (a nineteenth-century Moscow woman who specialized in triplets), which sounds stupendous until compared with the numbers achieved by some men to be mentioned below.

Hence multiple husbands do not help a woman to produce more babies, and very few human societies regularly practice polyandry. In the only such society that has received much study, the Tre-ba of Tibet, women with two husbands have on the average no more children than women with one husband. The reasons for Tre-ba polyandry are instead related to the Tre-ba system of land tenure: Tre-ba brothers often marry the same woman in order to avoid subdividing a small landholding.

Thus, a woman who "chooses" to care for her offspring is not thereby foreclosing other spectacular reproductive opportunities. In contrast, a polyandrous female phalarope produces on the average only 1.3 fledged chicks with one mate, but 2.2 chicks if she can corner two mates, and 3.7 chicks if she can corner three. A woman also differs in that respect from a man, whose theoretical ability to impregnate all the women of the world we have already discussed. Unlike the genetic unprofitability of polyandry for Tre-ba women, polygyny paid off well for nineteenth-century Mormon men, whose average lifetime output of children increased from a mere seven children for Mormon men with one wife to sixteen or twenty children for men with two or three wives, respectively, and to twenty-five children for Mormon church leaders, who averaged five wives.

Even these benefits of polygyny are modest compared to the hundreds of children sired by modern princes able to commandeer the resources of a centralized society for rearing their offspring without directly providing child care themselves. A nineteenth-century visitor to the court of the Nizam of Hyderabad, an Indian prince with an especially large harem, happened to be present during an eight-day period when four of the Nizam's wives gave birth, with nine more births anticipated for the following week. The record for lifetime number of offspring sired is credited to Morocco's Emperor Ismail the Bloodthirsty, father of seven hundred sons and an uncounted but presumably comparable number of daughters. These numbers make it clear that a man who fertilizes one woman and then devotes himself to child care may by that choice foreclose enormous alternative opportunities.

The remaining factor tending to make child care genetically less rewarding for men than for women is the justified paranoia about paternity that men share with the males of all other internally fertilized species. A man who opts for child care runs the risk that, unbeknownst to him, his efforts are transmitting the genes of a rival. This biological fact is the underlying cause for a host of repulsive practices by which men of various societies have sought to increase their confidence in paternity by restricting their wife's opportunity for sex with other men. Among such practices are high bride prices only for brides delivered as proven virgin goods; traditional adultery laws that define adultery by the marital status only of the participating woman (that of the participating man being irrelevant); chaperoning or virtual imprisonment of women; female "circumcision" (clitoridectomy) to reduce a woman's interest in initiating sex, whether marital or extramarital; and infibulation (suturing a woman's labia majora nearly shut so as to make intercourse impossible while the husband is away).

All three factors—sex differences in obligate parental investment, alternative opportunities foreclosed by child care, and confidence in parenthood—contribute to making men much more prone than women to desert a spouse and child. However, a man is not like a male hummingbird, male tiger, or the male of many other animal species, who can safely fly or walk away immediately after copulation, secure in the knowledge that his deserted female sex partner will be able to handle all the ensuing work of promoting the survival of his genes. Human infants virtually need biparental care, especially in traditional societies. While we shall see in chapter 5 that activities represented as male parental care may actually have more complex functions than meet the eye, many or most men in traditional societies do undoubtedly provide services to their children and spouse. Those services include: acquiring and delivering food; offering protection, not only against predators but also against other men who are sexually interested in a mother and regard her offspring (their potential stepchildren) as a competing genetic nuisance; owning land and making its produce available; building a house, clearing a garden, and performing other useful labor; and educating children, especially sons, so as to increase the children's chances of survival.

Sex differences in the genetic value of parental care to the parent provide a biological basis for the all-too-familiar differing attitudes of men and women toward extramarital sex. Because a human child virtually required paternal care in traditional human societies, extramarital sex is most profitable for a man if it is with a married woman whose husband will unknowingly rear the resulting child. Casual sex between a man and a married woman tends to increase the man's output of children, but not the woman's. That decisive difference is reflected in men's and women's differing motivations. Attitude surveys in a wide variety of human societies around the world have shown that men tend to be more interested than women in sexual variety, including casual sex and brief relationships. That attitude is readily understandable because it tends to maximize transmission of the genes of a man but not of a woman. In contrast, the motivation of a woman participating in extramarital sex is more often self-reported as marital dissatisfaction. Such a woman tends to be searching for a new lasting relationship: either a new marriage or a lengthy extramarital relationship with a man better able than her husband to provide resources or good genes.

....

CHAPTER 3

WHY DON'T MEN BREAST-FEED THEIR BABIES?

The Non-Evolution of Male Lactation

Today, we men are expected to share in the care of our children. We have no excuse not to, because we are perfectly capable of doing for our kids virtually anything that our wives can do. And so, when my twin sons were born in 1987, I duly learned to change diapers, clean up vomit, and perform the other tasks that come with parenthood.

The one task that I felt excused from was nursing my infants. It was visibly a tiring task for my wife. Friends kidded me that I should get hormone injections and share the burden. Yet cruel biological facts seemingly confront those who would bring sexual equality into this last bastion of female privilege or male cop-out. It appears obvious that males lack the anatomical equipment, the priming experience of pregnancy, and the hormones necessary for lactation. Until 1994, not a single one of the world's 4,300 mammal species was suspected of male lactation under normal conditions. The nonexistence of male lactation may thus seem to be a solved problem requiring no further discussion, and it may seem doubly irrelevant to a book about how the unique aspects of human sexuality evolved. After all, the problem's solution seems to depend on facts of physiology rather than on evolutionary reasoning, and exclusively female lactation is apparently a universal mammalian phenomenon not at all unique to humans.

In reality, the subject of male lactation follows perfectly from our discussion of the battle of the sexes. It illustrates the failure of strictly physiological explanations and the importance of evolutionary reasoning for understanding human sexuality. Yes, it's true that no male mammal has ever become pregnant, and that the great majority of male mammals normally don't lactate. But one has to go further and ask why mammals evolved genes specifying that only females, not males, would develop the necessary anatomical equipment, the priming experience of pregnancy, and the necessary hormones. Both male and female pigeons secrete crop "milk" to nurse their squab; why not men as well as women? Among seahorses it's the male rather than the female that becomes pregnant; why is that not also true for humans?

As for the supposed necessity of pregnancy as a primary experience for lactation, many female mammals, including many (most?) women, can produce milk without first being primed by pregnancy. Many male mammals, including some men, undergo breast development and lactate when given the appropriate hormones. Under certain conditions, a considerable fraction of men experience breast development and milk production even without having been treated hormonally. Cases of spontaneous lactation have long been known in male domestic goats, and the first case of male lactation in a wild mammal species has been reported recently.

Thus, lactation lies within the physiological potential of men. As we shall see, lactation would make more evolutionary sense for modern men than for males of most other mammal species. But the fact remains that it's not part of our normal repertoire, nor is it known to fall within the normal repertoire of other mammal species except for that single case reported recently. Since natural selection evidently could have made men lactate, why didn't it? That turns out to be a major question that cannot be answered simply by pointing to the deficiencies of male equipment. Male lactation beautifully illustrates all the main themes in the evolution of sexuality: evolutionary conflicts between males and females, the importance of confidence in paternity or maternity, differences in reproductive investment between the sexes, and a species' commitment to its biological inheritance.

As the first step in exploring these themes, I have to overcome your resistance to even thinking about male lactation, a product of our unquestioned assumption that it's physiologically impossible. The genetic differences between males and females, including those that normally reserve lactation for females, turn out to be slight and labile. This chapter will convince you of the feasibility of male lactation and will then explore why that theoretical possibility normally languishes unrealized.

•••••

Our sex is ultimately laid down by our genes, which in humans are bundled together in each body cell in twenty-three pairs of microscopic packages called chromosomes. One member of each of our twenty-three pairs was acquired from our mother, and the other member from our father. The twenty-three human chromosome pairs can be numbered and distinguished from each other by consistent differences in appearance. In chromosome pairs 1 through 22, the two members of each pair appear identical when viewed through a microscope. Only in the case of chromosome pair 23, the so-called sex chromosomes, do the two representatives differ, and even that's true only in men, who have a big chromosome (termed an X chromosome) paired with a small one (a Y chromosome). Women instead have two paired X chromosomes.

What do the sex chromosomes do? Many X chromosome

genes specify traits unrelated to sex, such as the ability to distinguish red and green colors. However, the Y chromosome contains genes specifying the development of testes. In the fifth week after fertilization human embryos of either sex develop a "bipotential" gonad that can become either a testis or an ovary. If a Y chromosome is present, that bet-hedging gonad begins to commit itself in the seventh week to becoming a testis, but if there's no Y chromosome, the gonad waits until the thirteenth week to develop as an ovary.

That may seem surprising: one might have expected the second X chromosome of girls to make ovaries, and the Y chromosome of boys to make testes. In fact, though, people abnormally endowed with one Y and two X chromosomes turn out most like males, whereas people endowed with three or just one X chromosome turn out most like females. Thus, the natural tendency of our bet-hedging primordial gonad is to develop as an ovary if nothing intervenes; something extra, a Y chromosome, is required to change it into a testis.

It's tempting to restate this simple fact in emotionally loaded terms. As the endocrinologist Alfred Jost put it, "Becoming a male is a prolonged, uneasy, and risky venture; it is a kind of struggle against inherent trends towards femaleness." Chauvinists might go further and hail becoming a man as heroic, and becoming a woman as the easy fallback position. Conversely, one might regard womanhood as the natural state of humanity, with men just a pathological aberration that regrettably must be tolerated as the price for making more women. I prefer merely to acknowledge that a Y chromosome switches gonad development from the ovarian path to the testicular path, and to draw no metaphysical conclusions.

·····

But there's more to a man than testes alone. A penis and prostate gland are among the many other obvious necessities of manhood, just as women need more than ovaries (for instance, it helps to have a vagina). It turns out that the embryo is endowed with other bipotential structures besides the primordial gonad. Unlike the primordial gonad, though, these other bipolar structures have a potential that is not directly specified by the Y chromosome. Instead, secretions produced by the testes themselves are what channel these other structures toward developing into male organs, while lack of testicular secretions channels them toward making female organs.

For example, already in the eighth week of gestation the testes begin producing the steroid hormone testosterone, some of which gets converted into the closely related steroid dihydrotestosterone. These steroids (known as androgens) convert some all-purpose embryonic structures into the glans penis, penis shaft, and scrotum; the same structures would otherwise develop into the clitoris, labia minora, and labia majora. Embryos also start out bet-hedging with two sets of ducts, known as the Mullerian ducts and Wolffian ducts. In the absence of testes, the Wolffian ducts atrophy, while the Mullerian ducts grow into a female fetus's uterus, fallopian tubes, and interior vagina. With testes present, the opposite happens: androgens stimulate the Wolffian ducts to grow into a male fetus's seminal vesicles, vas deferens, and epididymis. At the same time, a testicular protein called Mullerian inhibiting hormone does what its name implies: it prevents the Mullerian ducts from developing into the internal female organs.

Since a Y chromosome specifies testes, and since the presence or absence of the testes' secretions specifies the remaining male or female structures, it might seem as if there's no way that a developing human could end up with ambiguous sexual anatomy. Instead, you might think that a Y chromosome should guarantee 100 percent male organs, and that lack of a Y chromosome should guarantee 100 percent female organs.

In fact, a long series of biochemical steps is required to produce all those other structures besides ovaries or testes. Each step involves the synthesis of one molecular ingredient, termed an enzyme, specified by one gene. Any enzyme can be defective or absent if its underlying gene is altered by a mutation. Thus, an enzyme defect may result in a male pseudohermaphrodite, defined as someone possessing some female structures as well as testes. In a male pseudohermaphrodite with an enzyme defect, there is normal development of the male structures dependent on enzymes that act at the steps of the metabolic pathway before the defective enzyme. However, male structures dependent on the defective enzyme itself or on subsequent biochemical steps fail to develop and are replaced either by their female equivalent or by nothing at all. For example, one type of pseudohermaphrodite looks like a normal woman. Indeed, "she" conforms to the male ideal of female pulchritude even more closely than does the average real woman, because "her" breasts are well developed and "her" legs are long and graceful. Hence cases have turned up repeatedly of beautiful women fashion models not realizing that they are actually men with a single mutant gene until genetically tested as adults.

Since this type of pseudohermaphrodite looks like a normal girl baby at birth and undergoes externally normal development and puberty, the problem isn't even likely to be recognized until the adolescent "girl" consults a doctor over failure to begin menstruating. At that point, the doctor discovers a simple reason for that failure: the patient has no uterus, fallopian tubes, or upper vagina. Instead, the vagina ends blindly after two inches. Further examination reveals testes that secrete normal testosterone, are programmed by a normal Y chromosome, and are abnormal only for being buried in the groin or labia. In other words, the beautiful model is an otherwise normal male who happens to have a genetically determined biochemical block in his ability to respond to testosterone.

That block turns out to be in the cell receptor that would normally bind testosterone and dihydrotestosterone, thereby enabling those androgens to trigger the further developmental steps of the normal male. Since the Y chromosome is normal, the testes themselves form normally and produce normal Mullerian inhibiting hormone, which acts as in any man to forestall development of the uterus and fallopian tubes. However, development of the usual male machinery to respond to testosterone is interrupted. Hence development of the remaining bipotential embryonic sex organs follows the female channel by default: female rather than male external genitalia, and atrophy of the Wolffian ducts and hence of potential male internal genitalia. In fact, since the testes and adrenal glands secrete small amounts of estrogen that would normally be overridden by androgen receptors, the complete lack of those receptors in functional form (they are present in small numbers in normal women) makes the male pseudohermaphrodite appear externally superfeminine.

Thus, the overall genetic difference between men and women is modest, despite the big consequences of that modest difference. A small number of genes on chromosome 23, acting in concert with genes on other chromosomes, ultimately determine all differences between men and women. The differences, of course, include not just those in the reproductive organs themselves but also all other postadolescent sex-linked differences, such as the differences in beards, body hair, pitch of voice, and breast development.

• • • • •

The actual effects of testosterone and its chemical derivatives vary with age, organ, and species. Animal species differ greatly in how the sexes differ, and not only in mammary gland development. Even among higher anthropoids—humans and our closest relatives, the apes—there are familiar differences in sexual distinctiveness. We know from zoos and photos that adult male and female gorillas differ obviously at a long distance by the male's much greater size (his weight is double the female's), different shape of head, and silver-haired back. Men also differ, though much less obviously, from women in being slightly heavier (by 20 percent on the average), more muscular, and bearded. Even the degree of that difference varies among human populations: for example, the difference is less marked among Southeast Asians and Native Americans, since men of those populations have on the average much less body hair and beard development than in Europe and Southwest Asia. But males and females of some gibbon species look so similar that you couldn't distinguish them unless they permitted you to examine their genitals.

In particular, both sexes of placental mammals have mammary glands. While the glands are less well developed and nonfunctional in males of most mammal species, that degree of male underdevelopment varies among species. At the one extreme, in male mice and rats, the mammary tissue never forms ducts or a nipple and remains invisible from the outside. At the opposite extreme, in dogs and primates (including humans) the gland does form ducts and a nipple in both males and females and scarcely differs between the sexes before puberty.

During adolescence the visible differences between the mammalian sexes increase under the influence of a mix of hormones from the gonads, adrenal glands, and pituitary gland. Hormones released in pregnant and lactating females produce a further mammary growth spurt and start milk production, which is then reflexly stimulated by nursing. In humans, milk production is especially under the control of the hormone prolactin, while the responsible hormones in cows includes somatotropin, alias "growth hormone" (the hormone behind the current debate over proposed hormonal stimulation of milk cows).

It should be emphasized that male/female differences in hormones aren't absolute but a matter of degree: one sex may have higher concentrations and more receptors for a particular hormone. In particular, becoming pregnant is not the only way to acquire the hormones necessary for breast growth and milk production. For instance, normally circulating hormones stimulate a milk production, termed witch's milk, in newborns of several mammal species. Direct injection of the hormones estrogen or progesterone (normally released during pregnancy) triggers breast growth and milk production in virgin female cows and goats—and also in steers, male goats, and male guinea pigs. The hormonally treated virgin cows produced on the average as much milk as their half-sisters that were nursing calves to which they had given birth. Granted, hormonally treated steers produced much less milk than virgin cows; you shouldn't count on steer's milk in the supermarkets by next Christmas. But that's not surprising since the steers have previously limited their options: they haven't developed an udder to accommodate all the mammary gland tissue that hormonally treated virgin cows can accommodate.

There are numerous conditions under which injected or topically applied hormones have produced inappropriate breast development and milk secretion in humans, both in men and in nonpregnant or non-nursing women. Men and women cancer patients being treated with estrogen proceeded to secrete milk when injected with prolactin; among such patients was a sixty-four-year-old man who continued to produce milk for seven years after hormonal treatment was discontinued. (This observation was made in the 1940s, long before the regulation of medical research by human subjects protection committees, which now forbid such experiments). Inappropriate lactation has been observed in people taking tranquilizers that influence the hypothalamus (which controls the pituitary gland, the source of prolactin); it also has been observed in people recovering from surgery that stimulated nerves related to the suckling reflex, as well as in some women on prolonged courses of estrogen and progesterone birth-control pills. My favorite case is the chauvinist husband who kept complaining about his wife's "miserable little breasts," until he was shocked to find his own breasts growing. It turned out that his wife had been lavishly applying estrogen cream to her breasts to stimulate the growth craved by her husband, and the cream had been rubbing off on him.

·····

At this point, you may be starting to wonder whether all these examples are irrelevant to the possibility of normal male lactation, since they involve medical interventions such as hormone injections or surgery. But inappropriate lactation can occur without high-tech medical procedures: mere repeated mechanical stimulation of the nipples suffices to trigger milk secretion in virgin females of several mammal species, including humans. Mechanical stimulation is a natural way of releasing hormones by means of nerve reflexes connecting the nipples to hormone-releasing glands via the central nervous system. For instance, a sexually mature but virgin female marsupial can regularly be stimulated to lactate just by fostering another mother's young onto her teats. The "milking" of virgin female goats similarly triggers them to lactate. That principle might be transferable to men, since manual stimulation of the nipples causes a prolactin surge in men as well as in nonlactating women. Lactation is a not infrequent result of nipple self-stimulation in teenage boys.

My favorite human example of this phenomenon comes from a letter to the widely syndicated newspaper column "Dear Abby." An unmarried woman about to adopt a newborn infant longed to nurse the infant and asked Abby whether taking hormones would help her to do so. Abby's reply was: Preposterous, you'll only make yourself sprout hair! Several indignant readers then wrote in to describe cases of women in similar situations who succeeded in nursing an infant by repeatedly placing it at the breast.

Recent experience of physicians and nurse lactation specialists now suggest that most adoptive mothers can begin producing some milk within three or four weeks. The recommended preparation for prospective adoptive mothers is to use a breast pump every few hours to simulate sucking, beginning about a month before the expected delivery of the birth mother. Long before the advent of modern breast pumps, the same result was achieved by repeatedly putting a puppy or a human infant to the breast. Such preparation was practiced especially in traditional societies when a pregnant woman was sickly and her own mother wanted to be ready to step in and nurse the infant in case the daughter proved unable to do so. The reported examples include grandmothers up to the age of seventy-one, as well as Ruth's mother-in-law Naomi in the Old Testament. (If you don't believe it, open a Bible and turn to the Book of Ruth, chapter 4, verse 16.)

Breast development occurs commonly, and spontaneous lactation occasionally, in men recovering from starvation. Thousands of cases were recorded in prisoners of war released from concentration camps after World War II; one observer noted five hundred cases in survivors of one Japanese POW camp alone. The likely explanation is that starvation inhibits not only the glands that produce hormones but also the liver, which destroys those hormones. The glands recover much faster than the liver when normal nutrition is resumed, so that hormone levels soar unchecked. Again, turn to the Bible to discover how Old Testament patriarchs anticipated modern physiologists: Job (chapter 21, verse 24) remarked of a well-fed man that "His breasts are full of milk."

It has been known for a long time that many otherwise perfectly normal male goats, with normal testes and proven ability to inseminate females, surprise their owners by spontaneously growing udders and secreting milk. Billy-goat milk is similar in composition to she-goat milk but has even higher fat and protein content. Spontaneous lactation has also been observed in a captive monkey, the stump-tailed macaque of Southeast Asia.

In 1994, spontaneous male lactation was at last reported in males of a wild animal species, the Dyak fruit bat of Malaysia and adjacent islands. Eleven adult males captured alive proved to have functional mammary glands that yielded milk when manually expressed. Some of the males' mammary glands were distended with milk, suggesting that they had not been suckled and as a result milk had accumulated. However, others may have been suckled because they had less distended (but still functional) glands, as in lactating females. Among three samples of Dyak fruit bats caught at different places and seasons, two included lactating males, lactating females, and pregnant females, but adults of both sexes in the third sample were reproductively inactive. This suggests that male lactation in these bats may develop along with female lactation as part of the natural reproductive cycle. Microscopic examination of the testes revealed apparently normal sperm development in the lactating males.

Thus, while usually mothers lactate and fathers don't, males of at least some mammal species have much of the necessary anatomical equipment, physiological potential, and hormone receptors. Males treated either with the hormones themselves, or with other agents likely to release hormones, may undergo breast development and some lactation. There are several reports of apparently normal adult men nursing babies; one such man whose milk was analyzed secreted milk sugar, protein, and electrolytes at levels similar to those of mother's milk. All these facts suggest that it would have been easy for male lactation to evolve; perhaps it would have required just a few mutations causing increased release or decreased breakdown of hormones.

Evidently, evolution just didn't design men to utilize that physiological potential under normal conditions. In computing terminology, at least some males have the hardware; we merely haven't been programmed by natural selection to use it. Why not?

. . . . .

To understand why, we need to switch from physiological reasoning, which we have been using throughout this chapter, back to the evolutionary reasoning that we were using in chapter 2. In particular, recall how the evolutionary battle of the sexes has resulted in parental care being provided by the mother alone in about 90 percent of all mammal species. For those species, in which offspring will survive with zero paternal care, it's obvious that the question of male lactation never arises. Not only do males of those species have no need to lactate; they also don't have to bring food, defend a family territory, defend or teach their offspring, or do anything else for their offspring. The male's crass genetic interests are best served by chasing other females to impregnate. A noble male carrying a mutation to nurse his offspring (or to care for them in any other way) would quickly be outbred by selfish normal males that forewent lactation and thereby became able to sire more offspring.

Only for those 10 percent of mammal species in which male parental care is necessary does the question of male lactation even deserve consideration. Those minority species include lions, wolves, gibbons, marmosets—and humans. But even in those species requiring male parenting, lactation isn't necessarily the most valuable form that the father's contribution can take. What a big lion really must do is to drive off hyenas and other big lions bent on killing his cubs. He should be out patrolling his territory, not sitting home nursing the cubs (which the smaller lioness is perfectly capable of doing) while his cubs' enemies are sneaking up. The wolf father may make his most useful contribution by leaving the den to hunt, bringing back meat to the wolf mother, and letting her turn the meat into milk. The gibbon father may contribute best by looking out for pythons and eagles that might grab his offspring, and by vigilantly expelling other gibbons from the fruit trees in which his spouse and offspring are feeding, while marmoset fathers spend much time carrying their twin offspring.

All these excuses for male nonlactation still leave open the possibility that some other mammal species could exist in which male lactation might be advantageous to the male and his offspring. The Dyak fruit bat may turn out to be such a species. But even if there are mammal species for which male lactation would be advantageous, its realization runs up against problems posed by the phenomenon termed evolutionary commitment.

The idea behind evolutionary commitment can be understood by analogy to devices manufactured by humans. A manufacturer of trucks can easily modify one basic truck model for different but related purposes, such as transporting furniture, horses, or frozen food. Those different purposes can be fulfilled by making a few minor variations on the same basic design of the truck's cargo compartment, with little or no change in the motor, brakes, axles, and other major components. Similarly, an airplane manufacturer can with minor modifications use the same model of airplane to carry ordinary passengers, skydivers, or freight. But it is not feasible to convert a truck into an airplane or vice versa, because a truck is committed to truckhood in too many respects: heavy body, diesel motor, braking system, axles, and so on. To build an airplane, one would not start with a truck and modify it; one would instead start all over again.

Animals, in contrast, are not designed from scratch to provide an optimal solution for a desired lifestyle. Instead, they evolve from existing animal populations. Evolutionary changes in lifestyle come about incrementally through the accumulation of small changes in an evolutionary design adapted to a different but related lifestyle. An animal with many adaptations to one specialized lifestyle may not be able to evolve the many adaptations required for a different lifestyle, or may do so only after a very long time. For instance, a female mammal that gives birth to live young cannot evolve into a birdlike egg layer merely by extruding her embryo to the outside within a day of fertilization; she would have to have evolved birdlike mechanisms for synthesizing yolk, eggshell, and other avian commitments to egg laying.

Recall that, of the two main classes of warm-blooded vertebrates, birds and mammals, male parental care is the rule among birds and the exception among mammals. That difference results from birds' and mammals' long evolutionary histories of developing different solutions to the problem of what to do with an egg that has just been fertilized internally. Each of those solutions has required a whole set of adaptations, which differ between birds and mammals and to which all modern birds and mammals are now heavily committed.

The bird's solution is to have the female rapidly extrude the fertilized embryo, packaged with yolk inside a hard shell, in an extremely undeveloped and utterly helpless state that is impossible for anyone except an embryologist to recognize as a bird. From the moment of fertilization to the moment of extrusion, the embryo's development inside the mother lasts only a day or a few days. That brief internal development is followed by a much longer period of development outside the mother's body: up to 80 days of incubation before the egg hatches, and up to 240 days of feeding and caring for the hatched chick until it can fly.

Once the egg has been laid, there is nothing further in the chick's development that uniquely requires its mother's help. The father can sit on the egg and keep it warm just as well as the mother can. After hatching out, chicks of most bird species eat the same food as their parents, and the father can collect and bring that food to the nest as well as the mother can.

In most bird species the care of the nest, egg, and chick requires both parents. In those bird species in which the efforts of one parent suffice, that parent is more often the mother than the father, for the reasons discussed in chapter 2: the female's greater obligate internal investment in the fertilized embryo, the greater opportunities foreclosed for the male by parental care, and the male's low confidence in paternity as a result of internal fertilization. But in all bird species the female's obligate internal investment is much less than that in any mammal species, because the developing young bird is "born" (laid) in such an early stage of development compared to even the least developed newborn mammal. The ratio of development time outside the mother—a time of duties that in theory can be shared by the mother and the father—to development time inside the mother is much higher for birds than for mammals. No mother bird's "pregnancy"—egg formation time—approaches the nine months of human pregnancy or even the twelve days of the briefest mammalian pregnancy.

Hence female birds are not as easily bluffed as female mammals into caring for the offspring while the father deserts to philander. That has consequences for the evolutionary programming not only of birds' instinctive behaviors but also of their anatomy and physiology. In pigeons, which feed their young by secreting "milk" from their crops, both the father and the mother have evolved to secrete milk. Biparental care is the rule in birds, and while in those bird species that practice uniparental care the mother is usually the sole caretaker, in some bird species it is the father, a development unprecedented among mammals. Care by the father alone characterizes not only those bird species characterized by sex-role-reversal polyandry but also some other birds, including ostriches, emus, and tinamous.

The bird solution to the problems posed by internal fertilization and subsequent embryonic development involves specialized anatomy and physiology. Female but not male birds possess an oviduct of which one portion secretes albumin (the egg white protein), another portion makes the inner and outer shell membranes, and still another makes the eggshell itself. All of those hormonally regulated structures and their metabolic machinery represent evolutionary commitment. Birds must have been evolving along this pathway for a long time, because egg laying was already widespread in ancestral reptiles, from which birds may have inherited much of their egg-making machinery. Creatures that are recognizably birds and no longer reptiles, such as the famous Archaeopteryx, appear in the fossil record by 150 million years ago. While the reproductive biology of Archaeopteryx is unknown, a dinosaur fossil from about 80 million years ago has been found entombed on a nest and eggs, suggesting that birds inherited nesting behavior as well as egg laying from their reptilian ancestors.

Modern bird species vary greatly in their ecology and lifestyle, from aerial fliers to terrestrial runners and marine divers, from tiny hummingbirds to giant extinct elephant birds, and from penguins nesting in the Antarctic winter to toucans breeding in tropical rainforests. Despite that variation in lifestyle, all existing birds have remained committed to internal fertilization, egg laying, incubation, and other distinctive features of avian reproductive biology, with only minor variations among species. (The principal exceptions are the brush turkeys of Australia and the Pacific islands: they incubate their eggs with external heat sources, such as fermentative, volcanic, or solar heat, rather than with body heat.) If one were designing a bird from scratch, perhaps one could come up with a better but entirely different reproductive strategy, such as that of bats, which fly like birds but reproduce by pregnancy, live birth, and lactation. Whatever the virtues of that bat solution, it would require too many major changes for birds, which remain committed to their own solution.

•••••

Mammals have their own long history of evolutionary commitment to their solution to the same problem of what to do with an internally fertilized egg. The mammalian solution begins with pregnancy, an obligate period of embryonic development within the mother that lasts much longer than in any mother bird. Pregnancy's duration ranges from a minimum of twelve days in bandicoots to twenty-two months in elephants. That big initial commitment by a female mammal makes it impossible for her to bluff her way out of further commitment and has led to the evolution of female lactation. Like birds, mammals have evidently been committed to their distinctive solution for a long time. Lactation does not leave fossil traces, but it is shared among the three living groups of mammals (monotremes, marsupials, and placentals), which had already differentiated from each other by 135 million years ago. Hence lactation presumably arose in some mammal-like reptilian ancestor (so-called therapsid reptiles) even earlier.

Like birds, mammals are committed to much specialized reproductive anatomy and physiology of their own. Some of those specializations differ greatly between the three mammalian groups, such as placental development resulting in a relatively mature newborn in placental mammals, earlier birth and relatively longer postnatal development in marsupials, and egg-laying in monotremes. These specializations have probably been in place for at least 135 million years.

Compared to those differences between the three mammalian groups, or compared to the differences between all mammals and birds, variation within each of the three groups of mammals is minor. No mammal has re-evolved external fertilization or discarded lactation. No marsupial or placental mammal has re-evolved egg laying. Species differences in lactation are mere quantitative differences: more of this, less of that. For instance, the milk of Arctic seals is concentrated in nutrients, high in fat, and almost devoid of sugar, while human milk is more dilute in nutrients, sugary, and low in fat. Weaning from milk to solid food extends over a period of up to four years in traditional human hunter-gatherer societies. At the other extreme, guinea pigs and jackrabbits are capable of nibbling solid food within a few days of birth and dispensing with milk soon thereafter. Guinea pigs and jackrabbits may be evolving in the direction of bird species with precocial young, such as chickens and shorebirds, whose hatchlings already have open eyes, can run, and can find their own food but cannot yet fly or fully regulate their own body temperature. Perhaps, if life on Earth survives the current onslaught by humans, the evolutionary descendants of guinea pigs and jackrabbits will discard their inherited evolutionary commitment to lactation—in a few more tens of millions of years.

Thus, other reproductive strategies might work for a mammal, and it would seem to require few mutations to transform a newborn guinea pig or jackrabbit into a newborn mammal that requires no milk at all. But that has not happened: mammals have remained evolutionarily committed to their characteristic reproductive strategy. Similarly, even though we have seen that male lactation is physiologically possible, and although it also would seem to require few mutations, female mammals have nevertheless had an enormous evolutionary head start on males in perfecting their shared physiological potential for lactation. Females, but not males, have been undergoing natural selection for milk production for tens of millions of years. In all the species I cited to demonstrate that male lactation is physiologically possible—humans, cows, goats, dogs, guinea pigs, and Dyak fruit bats—lactating males still produce much less milk than do females.

· · · · ·

Still, the tantalizing recent discoveries about Dyak fruit bats make one wonder whether out there today, undiscovered, might be some mammal species whose males and females share the burden of lactation—or one that might evolve such sharing in the future. The life history of the Dyak fruit bat remains virtually unknown, so we cannot say what conditions favored in it the beginnings of normal male lactation, nor how much milk (if any) the male bats actually supply to their offspring. Nevertheless, we can easily predict on theoretical grounds the conditions that would favor the evolution of normal male lactation. Those conditions include: a litter of infants that constitute a big burden to nourish; monogamous male-female pairs; high confidence of males in their paternity; and hormonal preparation of fathers, while their mate is still pregnant, for eventual lactation.

The mammal species that some of these conditions already best describe is—the human species. Medical technology is making others of these conditions increasingly applicable to us. With modern fertility drugs and high-tech methods of fertilization, births of twins and triplets are becoming more frequent. Nursing human twins is such an energy drain that the daily energy budget of a mother of twins approaches that of a soldier in boot camp. Despite all our jokes about infidelity, genetic testing shows the great majority of American and European babies tested to have been actually sired by the mother's husband. Genetic testing of fetuses is becoming increasingly common and can already permit a man to be virtually 100 percent sure that he really sired the fetus within his pregnant wife.

Among animals, external fertilization favors, and internal fertilization mitigates against, the evolution of male parental investment. That fact has discouraged male parental investment by other mammal species but now uniquely favors it in humans, because in-vitro external fertilization techniques have become a reality for humans within the past two decades. Of course, the vast majority of the world's babies are still conceived internally by natural methods. But the increasing number of older women and men who wish to conceive but have difficulty doing so, and the reported modern decline in human fertility (if it is real), combine to ensure that more and more human babies will be products of external fertilization, like most fish and frogs.

All these features make the human species a leading candidate for male lactation. While that candidacy may take millions of years to perfect through natural selection, we have it in our power to short-circuit that evolutionary process by technology. Some combination of manual nipple stimulation and hormone injections may soon develop the latent potential of the expectant father—his confidence in paternity buttressed by DNA testing—to make milk, without the need to await genetic changes. The potential advantages of male lactation are numerous. It would promote a type of emotional bonding of father to child now available only to women. Many men, in fact, are jealous of the special bond arising from breast-feeding, whose traditional restriction to mothers makes men feel excluded. Today, many or most mothers in first-world societies have already become unavailable for breast-feeding, whether because of jobs, illness, or lactational failure. Yet not only parents but also babies derive many benefits from breast-feeding. Breast-fed babies acquire stronger immune defenses and are less susceptible to numerous diseases, including diarrhea, ear infections, early-age-onset diabetes, influenza, necrotizing enterocolitis, and SIDS (Sudden Infant Death Syndrome). Male lactation could provide those benefits to babies if the mother is unavailable for any reason.

It must be acknowledged, however, that the obstacles to male lactation are not only physiological ones, which can evidently be overcome, but also psychological ones. Men have traditionally regarded breast-feeding as a woman's job, and the first men to breast-feed their infants will undoubtedly be ridiculed by many other men. Nevertheless, human reproduction already involves increasing use of other procedures that would have seemed ridiculous until a few decades ago: procedures such as external fertilization without intercourse, fertilization of women over the age of fifty, gestation of one woman's fetus inside another woman's womb, and survival of prematurely delivered one-kilogram fetuses by high-tech incubator methods. We now know that our evolutionary commitment to female lactation is physiologically labile; it may prove psychologically labile as well. Perhaps our greatest distinction as a species is our capacity, unique among animals, to make counter-evolutionary choices. Most of us choose to renounce murder, rape, and genocide, despite their advantages as a means for transmitting our genes, and despite their widespread occurrence among other animal species and earlier human societies. Will male lactation become another such counter-evolutionary choice?

....

CHAPTER 4

WRONG TIME FOR LOVE

The Evolution of Recreational Sex

First scene: a dimly lit bedroom, with a handsome man lying in bed. A beautiful young woman in a nightgown runs to the bed. A diamond wedding ring flashes virtuously on her left hand, while her right hand clutches a small blue strip of paper. She bends down and kisses the man's ear.

She: "Darling! It's *exactly* the right time!"

Next scene: same bedroom, same couple, evidently making love, but details tastefully obscured by the dim lighting. Then the camera shifts to a calendar slowly being flipped (to indicate the passage of time) by a graceful hand wearing the same diamond wedding ring.

Next scene: the same beautiful couple, blissfully holding a clean smiling baby.

He: "Darling! I'm so glad that Ovu-stick told us when it was *exactly* the right time!"

Last frame: close-up of the same graceful hand, clutching the small blue strip of paper. Caption reads: "Ovu-stick. Home urine test to detect ovulation."

If baboons could understand our TV ads, they'd find that one especially hilarious. Neither a male nor female baboon needs a hormonal test kit to detect the female's ovulation, the sole time when her ovary releases an egg and when she can be fertilized. Instead, the skin around the female's vagina swells and turns a bright red color visible at a distance. She also gives off a distinctive smell. In case a dumb male still misses the point, she crouches in front of him and presents her hindquarters. Most other female animals are equally aware of their own ovulation and advertise it to males with equally bold visual signals, odors, or behaviors.

We consider female baboons with bright red hindquarters bizarre. In fact, we humans are the ones whose scarcely detectable ovulations make us members of a small minority in the animal world. Men have no reliable means of detecting when their partners can be fertilized, nor did women in traditional societies. I grant that many women experience headaches or other sensations around the midpoint of a menstrual cycle. However, they wouldn't know that these are signs of ovulation if they hadn't been told so by scientists—and even scientists didn't figure that out until around 1930. Similarly, women can be *taught* to detect ovulation by monitoring their body temperature or mucus, but that's very different from the instinctive knowledge possessed by female animals. If we too had such instinctive knowledge, manufacturers of ovulation test kits and contraceptives wouldn't be doing such a booming business.

We're also bizarre in our nearly continuous practice of sex, a behavior that is a direct consequence of our concealed ovulations. Most other animal species confine sex to a brief estrous period around the advertised time of ovulation. (The noun *estrus* and adjective *estrous* are derived from the Greek word for "gadfly," an insect that pursues cattle and drives them into a frenzy.) At estrus, a female baboon emerges from a month of sexual abstinence to copulate up to one hundred times, while a female Barbary macaque does it on the average every seventeen minutes, distributing her favors at least once to every adult male in her troop. Monogamous gibbon couples go several years without sex, until the female weans her most recent infant and comes into estrus again. The gibbons relapse once more into abstinence as soon as the female becomes pregnant.

We humans, though, practice sex on any day of the estrus cycle. Women solicit it on any day, and men perform without being choosy about whether their partner is fertile or ovulating. After decades of scientific inquiry, it isn't even certain at what stage in the cycle a woman is most interested in men's sexual advances—if indeed her interest shows any cyclical variation. Hence most human copulations involve women who are unable to conceive at that moment. Not only do we have sex at the "wrong" time of the cycle, but we continue to have sex during pregnancy and after menopause, when we know for sure that fertilization is impossible. Many of my New Guinea friends feel obliged to have regular sex right up to the end of pregnancy, because they believe that repeated infusions of semen furnish the material to build the fetus's body.

Human sex does seem a monumental waste of effort from a "biological" point of view—if one follows Catholic dogma in equating sex's biological function with fertilization. Why don't women give clear ovulatory signals, like most other female animals, so that we can restrict sex to moments when it could do us some good? This chapter seeks to understand the evolution of concealed ovulation, nearly constant female sexual receptivity, and recreational sex—a trinity of bizarre reproductive behaviors that is central to human sexuality.

•••••

By now, you may have decided that I'm the prime example of an ivory tower scientist searching unnecessarily for problems to explain. I can hear several billion of the world's people protesting, "There's no problem to explain, except why Jared Diamond is such an idiot. *You* don't understand why *we* have sex all the time? Because it's fun, of course!"

Unfortunately, that answer doesn't satisfy scientists. While animals are engaged in sex, they too look as if they're having fun, to judge by their intense involvement. Marsupial mice even seem to be having lots more fun than we do, if the duration of their copulations (up to twelve hours) is any indication. Then why do most animals consider sex fun only when the female can be fertilized? Behavior evolves through natural selection, just as anatomy does. Hence if sex is enjoyable, natural selection must have been responsible for that outcome. Yes, sex is fun for dogs too, but only at the right time: dogs, like most other animals, have evolved the good sense to enjoy sex when it can do some good. Natural selection favors those individuals whose behavior lets them pass their genes to the most babies. How does it help you make more babies if you are crazy enough to enjoy sex at a time when you couldn't possibly make a baby?

A simple example illustrating the goal-directed nature of sexual activity in most animal species is provided by Pied Flycatchers, the bird species I discussed in chapter 2. Normally, a female Pied Flycatcher solicits copulation only when her eggs are ready to be fertilized, a few days before laying. Once she begins egg laying, her interest in sex vanishes and she resists propositions from males or behaves indifferently toward them. But in an experiment in which a team of ornithologists made twenty female Pied Flycatchers into widows after completion of egg laying by removing their mates, six of the twenty experimental widows were seen to solicit copulation from new males within two days, three were seen actually to copulate, and more may have done so unobserved. Evidently, the females were attempting to trick the males into believing them to be fertile and available. When the eggs eventually hatched, the males would have no way of realizing that some other male had actually fathered the clutches. In at least a few cases, the trick worked, and the males proceeded to feed the hatchlings as a biological father would have. There was thus not the slightest indication that any of the females was a merry widow, pursuing sex for mere pleasure.

Since we humans are exceptional in our concealed ovulations, unceasing receptivity, and recreational sex, it can only be because we evolved to be that way. It's especially paradoxical that in *Homo sapiens*, the species unique in its self-consciousness, females should be unconscious of their own ovulation, when female animals as dumb as cows are aware of it. Something special was required to conceal ovulation from a female as smart and aware as a woman. As we'll discover, it has proven unexpectedly difficult for scientists to figure out what that special something was.

There's a simple reason why most other animals are sensibly stingy about copulatory effort: sex is costly in energy, time, and risk of injury or death. Let me count the reasons why you should not love your beloved unnecessarily:

1: Sperm production is sufficiently costly for males that worms with a mutation that reduces sperm production live longer than normal worms.

2: Sex takes time that could otherwise be devoted to finding food.

3: Couples locked in embrace risk being surprised and killed by a predator or enemy.

4: Older individuals may succumb to the strain of sex: France's Emperor Napoleon the Third suffered a stroke while engaged in the act, and Nelson Rockefeller died during sex.

5: Fights between male animals competing for an estrous female often result in serious injury to the female as well as to the males.

6: Being caught at extramarital sex is risky for many animal species, including (most notoriously) humans.

Thus, we would reap a big advantage by being as sexually efficient as other animals. What compensating advantage do we get from our apparent inefficiency?

Scientific speculation tends to center on another of our unusual features: the helpless condition of human infants makes lots of parental care necessary for many years. The young of most mammals start to get their own food as soon as they're weaned; they become fully independent soon afterwards. Hence most female mammals can and do rear their young with no assistance from the father, whom the mother sees only to copulate. For humans, though, most food is acquired by complex technologies far beyond the dexterity or mental ability of a toddler. As a result, our children have to have food brought to them for at least a decade after weaning, and that job is much easier for two parents than for one. Even today it's hard for the single human mother to rear kids unassisted, and it used to be much harder in prehistoric days when we were hunter-gatherers.

Now consider the dilemma facing an ovulating cavewoman who has just been fertilized. In any other mammal species, the male who did it would promptly go off in search of another ovulating female to fertilize. For the cavewoman, though, the male's departure would expose her eventual child to the likelihood of starvation or murder. What can she do to keep that man? Her brilliant solution: remain sexually receptive even after ovulating! Keep him satisfied by copulating whenever he wants! In that way, he'll hang around, have no need to look for new sex partners, and will even share his daily hunting bag of meat. Recreational sex is thus supposed to function as the glue holding a human couple together while they cooperate in rearing their helpless baby. That in essence is the theory formerly accepted by anthropologists, and it seemed to have much to recommend it.

However, as we have learned more about animal behavior, we have come to realize that this sex-to-promote-family-values theory leaves many questions unanswered. Chimpanzees and especially bonobos have sex even more often than we do (as much as several times daily), yet they are promiscuous and have no pair-bond to maintain. Conversely, one can point to males of numerous mammal species that require no such sexual bribes to induce them to remain with their mate and offspring. Gibbons, which actually often live as monogamous couples, go years without sex. You can watch outside your window how male songbirds cooperate assiduously with their mates in feeding the nestlings, although sex ceased after fertilization. Even male gorillas with a harem of several females get only a few sexual opportunities each year; their mates are usually nursing or out of estrus. Why do women have to offer the sop of constant sex, when these other females don't?

There's a crucial difference between our human couples and those abstinent couples of other animal species. Gibbons, most songbirds, and gorillas live dispersed over the landscape, with each couple (or harem) occupying its separate territory. That pattern provides few encounters with potential extramarital sex partners. Perhaps the most distinctive feature of traditional human society is that mated couples live within large groups of other couples with whom they have to cooperate economically. To find an animal with parallel living arrangements, one has to go far beyond our mammalian relatives to densely packed colonies of nesting seabirds. Even seabird couples, though, aren't as dependent on each other economically as we are.

The human sexual dilemma, then, is that a father and mother must work together for years to rear their helpless children, despite being frequently tempted by other fertile adults nearby. The specter of marital disruption by extramarital sex, with its potentially disastrous consequences for parental cooperation in child-rearing, is pervasive in human societies. Somehow, we evolved concealed ovulation and constant receptivity to make possible our unique combination of marriage, coparenting, and adulterous temptation. How does it all fit together?

.....

Scientists' belated appreciation of these paradoxes has spawned an avalanche of competing theories, each of which tends to reflect the gender of its author. For instance, there's the prostitution theory proposed by a male scientist: women evolved to trade sexual favors for donations of meat from male hunters. There's also a male scientist's better-genes-through-cuckoldry theory, which reasons that a cavewoman with the misfortune to have been married off by her clan to an ineffectual husband could use her constant receptivity to attract (and be extramaritally impregnated by) a neighboring caveman with superior genes.

Then again, there's the anticontraceptive theory proposed by a woman scientist, who was well aware that childbirth is uniquely painful and dangerous in the human species because of the large size of the newborn human infant relative to its mother as compared to that ratio in our ape relatives. A one-hundred-pound woman typically gives birth to a six-pound infant, while a female gorilla twice that size (two hundred pounds) gives birth to an infant only half as large (three pounds). As a result, human mothers often died in childbirth before the advent of modern medical care, and women are still attended at birth by helpers (obstetricians and nurses in modern first-world societies, midwives or older women in traditional societies), whereas female gorillas give birth unattended and have never been recorded as dying in childbirth. Hence according to the anticontraceptive theory, cavewomen aware of the pain and danger of childbirth, and also aware of their day of ovulation, misused that knowledge to avoid sex then. Such women failed to pass on their genes, leaving the world populated by women ignorant of their time of ovulation and thus unable to avoid having sex while fertile.

From this plethora of hypotheses to explain concealed ovulation, two, which I shall refer to as the "daddy-at-home" theory and the "many-fathers" theory, have survived as most plausible. Interestingly, the two hypotheses are virtually opposite. The daddy-at-home theory posits that concealed ovulation evolved to promote monogamy, to force the man to stay home, and thus to bolster his certainty about his paternity of his wife's children. The many-fathers theory instead posits that concealed ovulation evolved to give the woman access to many sex partners and thus to leave many men uncertain as to whether they sired her children.

Take first the daddy-at-home theory, developed by the biologists Richard Alexander and Katharine Noonan of the University of Michigan. To understand their theory, imagine what married life would be like if women *did* advertise their ovulations, like female baboons with bright red derrières. A husband would infallibly recognize, from the color of his wife's derrière, the day on which she was ovulating. On that day he would stay home and assiduously make love in order to fertilize her and pass on his genes. On all other days, he would realize from his wife's pallid derrière that lovemaking with her was useless. He would instead wander off in search of other, unguarded, red-hued ladies, so that he could fertilize them too and pass on even more of his genes. He'd feel secure in leaving his wife at home then, because he'd know that she wasn't sexually receptive to men and couldn't be fertilized anyway. That's what male geese, seagulls, and Pied Flycatchers actually do.

For humans, the results of those marriages with advertised ovulations would be awful. Fathers would rarely be at home, mothers would be unable to rear kids unassisted, and babies would die in droves. That would be bad for both mothers *and* fathers, because neither would succeed in propagating their genes.

Now let's picture the reverse scenario, in which a husband has no clue to his wife's fertile days. He then has to stay at home and make love with her on as many days of the month as possible if he wants to have much chance of fertilizing her. Another motive for him to stay at home is to guard her constantly against other men, since she might prove to be fertile on any day that he is away. If the philandering husband has the bad luck to be in bed with another woman on the night when his wife happens to be ovulating, some other man might be in the philanderer's bed fertilizing his wife, while the philanderer himself is wasting his adulterous sperm on another woman unlikely to be ovulating then anyway. Under this reverse scenario, a man has less reason to wander, since he can't identify which of his neighbor's wives are fertile. The heartwarming outcome: fathers hang around and share baby care, with the result that babies survive. That's good for mothers as well as fathers, both of whom now succeed in transmitting their genes.

In effect, Alexander and Noonan argue that the peculiar physiology of the human female forces husbands to stay at home (at least, more than they would otherwise). The woman gains by recruiting an active coparent. But the man also gains, *provided* that he cooperates and plays by the rules of his wife's body. By staying home, he acquires confidence that the child whom he is helping to rear really does carry his genes. He needn't be fearful that, while he is off hunting, his wife (like a female baboon) may start flashing a bright red derrière as an advertisement for her imminent ovulation, thereby attracting swarms of suitors and publicly mating with every man around. Men accept these ground rules to such a degree that they continue to have sex with their wives during pregnancy and after menopause, when even men know that fertilization is impossible. Thus, in Alexander and Noonan's view, women's concealed ovulations and constant receptivity evolved in order to promote monogamy, paternal care, and fathers' confidence in their paternity.

Competing with this view is the many-fathers theory developed by the anthropologist Sarah Hrdy of the University of California at Davis. Anthropologists have long recognized that infanticide used to be common in many traditional human societies, although modern states now have laws against it. Until recent field studies by Hrdy and others, though, zoologists had no appreciation for how often infanticide occurs among animals as well. The species in which it has been documented now include our closest animal relatives, chimpanzees and gorillas, in addition to a wide range of other species from lions to African hunting dogs. Infanticide is especially likely to be committed by adult males against infants of females with whom they have never copulated—for example, when intruding males try to supplant resident males and acquire their harem of females. The usurper thus "knows" that the infants killed are not his own.

Naturally, infanticide horrifies us and makes us ask why animals (and formerly humans) do it so often. On reflection, one can see that the murderer gains a grisly genetic advantage. A female is unlikely to ovulate as long as she is nursing an infant. But a murderous intruder is genetically unrelated to the infants of a troop that he has just taken over. By killing such an infant, he terminates its mother's lactation and stimulates her to resume estrus cycles. In many or most cases of animal infanticide and takeovers, the murderer proceeds to fertilize the bereaved mother, who bears an infant carrying the murderer's own genes.

As a major cause of infant death, infanticide is a serious evolutionary problem for animal mothers, who thereby lose their genetic investment in murdered offspring. For in- stance, a typical female gorilla over the course of her life-time loses at least one of her offspring to infanticidal intruding male gorillas attempting to take over the harem to which she belongs. Indeed, over one-third of all infant gorilla deaths are due to infanticide. If a female has only a brief, conspicuously advertised estrus, a dominant male can easily monopolize her during that time. All other males consequently "know" that the resulting infant was sired by their rival, and they have no compunctions about killing the infant.

Suppose, though, that the female has concealed ovula-tions and constant sexual receptivity. She can exploit those advantages to copulate with many males—even if she has to do it sneakily, when her consort isn't looking. While no male can then be confident of his paternity, many males recognize that they *might* have sired the mother's eventual infant. If such a male later succeeds in driving out the mother's consort and taking her over, he avoids killing her infant because it could be his own. He might even help the infant with protection and other forms of paternal care. The mother's concealed ovulation will also serve to de-crease fighting between adult males within her troop be-cause any single copulation is unlikely to result in conception and hence is no longer worth fighting over.

As an example of how widely females may thus use concealed ovulation to confuse paternity, consider the African monkeys called vervets, familiar to anyone who has visited an East African game park. Vervets live in troops consisting of up to seven adult males and ten adult females. Since female vervets give no anatomical or behav-ioral signs of ovulation, the biologist Sandy Andelman sought out an acacia tree with a troop of vervets, stood under the tree, held up a funnel and bottle, collected urine when a female relieved herself, and analyzed the urine for hormonal signs of ovulation. Andelman also kept track of copulations. It turned out that females started to copulate long before they ovulated, continued long after they ovulated, and did not reach their peak sexual receptivity until the first half of pregnancy.

At that time the female's belly was not yet visibly bulging, and the deceived males had no idea that they were utterly wasting their efforts. Females finally ceased copulating during the latter half of pregnancy, when the males could no longer be deceived. That still left most males in the troop ample time to have sex with most of the troop's females. One-third of the males were able to copulate with every single female. Thus, through concealed ovulation female vervets ensured the benevolent neutrality of almost all of the potentially murderous males in their immediate neighborhood.

In short, Hrdy considers concealed ovulation an evolutionary adaptation by females to minimize the big threat to their offsprings' survival posed by adult males. Whereas Alexander and Noonan view concealed ovulation as clarifying paternity and reinforcing monogamy, Hrdy sees it as confusing paternity and effectively undoing monogamy.

At this point, you may be starting to wonder about a potential complication in both the daddy-at-home theory and the many-fathers theory. Why is human ovulation concealed from women as well, when all that's required by either theory is for women to conceal ovulation from men? For example, why couldn't women keep their derrières the same shade of red every day of the month to deceive men, while still remaining aware of sensations of ovulation and just faking an interest in sex with lusty men on non-ovulatory days?

The answer to that objection should be obvious: it

would be hard for a woman convincingly to fake sexual receptivity if she felt turned off and knew that she was currently infertile. That point applies with particular force to the daddy-at-home theory. When a woman is involved in a long-lasting monogamous relationship in which the partners come to know each other intimately, it would be hard for her to deceive her husband unless she herself were deceived as well.

There is no question that the many-fathers theory is plausible for those animal species (and perhaps those traditional human societies) in which infanticide is a big problem. But the theory seems hard to reconcile with modern human society as we know it. Yes, extramarital sex occurs, but doubts about paternity remain the exception, not the rule that drives society. Genetic tests show that at least 70 percent, perhaps even 95 percent, of American and British babies really are sired legitimately, that is, by the mother's husband. It's hardly the case that for each kid there are many men standing around radiating benevolent interest, or even showering gifts and dispensing protection, while thinking, "*I* may be that kid's *real* father!"

It therefore seems unlikely that protecting kids against infanticide is what propels women's constant sexual receptivity today. Nevertheless, as we'll now see, women may have had this motivation in our distant past, and sex may have subsequently assumed a different function that now sustains it.

· · · · ·

How, then, are we to evaluate these two competing theories? Like so many other questions about human evolution, this one can't be settled in the way preferred by chemists and molecular biologists, a test-tube experiment. Yes, we'd have a decisive test if there were some human population whose women we could cause to turn bright red at estrus and to remain frigid at other times, and whose men we could cause to be turned on only by bright red women. We could then see whether the result was more philandering and less paternal care (as predicted by the daddy-at-home theory) or less philandering and more infanticide (as predicted by the many-fathers theory). Alas for science, such a test is presently impossible, and it will remain immoral even if genetic engineering ever makes it possible.

But we can still resort to another powerful technique preferred by evolutionary biologists for solving such problems. It's termed the comparative method. We humans, it turns out, aren't unique in our concealment of ovulation. While it's exceptional among mammals in general, it's fairly common among higher primates (monkeys and apes), the group of mammals to which we belong. Dozens of primate species show no externally visible signs of ovulation; many others do show signs, albeit slight ones; and still others advertise it flagrantly. The reproductive biology of each species represents the outcome of an experiment, performed by nature, on the benefits and drawbacks of concealing ovulation. By comparing primate species, we can learn which features are shared by those species with concealed ovulation but are absent from those species with advertised ovulation.

That comparison throws new light on our sexual habits. It was the subject of an important study by the Swedish biologists Birgitta Sillén-Tullberg and Anders Møller. Their analysis proceeded in four steps.

Step 1. For as many higher primate species as possible (sixty-eight in all), Sillén-Tullberg and Møller tabulated visible signs of ovulation. Aha!—you may object immediately—visible to whom? A monkey may give signals invisible to us humans but obvious to another monkey, such as odors (pheromones). For example, cattle breeders trying to perform artificial insemination on a prize dairy cow have big problems figuring out when the cow is ovulating. Bulls, though, can tell easily by the cow's smell and behavior.

Yes, that problem can't be ignored, but it's more serious for cows than for higher primates. Most primates resemble us in being active by day, sleeping at night, and depending heavily on their eyes. A male rhesus monkey whose nose isn't working can still recognize an ovulating female monkey by the slight reddening around her vagina, even though her reddening is not nearly so obvious as in a female baboon. For those monkey species that we humans classify as having no visible signs of ovulation, it's often clear that the male monkeys are equally confused, because they copulate at totally inappropriate times, such as with non-estrous or pregnant females. Hence our own ratings of "visible signs" aren't worthless.

The result of this first step of the analysis was that nearly half of the primates studied—thirty-two out of sixty-eight—resemble humans in lacking visible signs of ovulation. Those thirty-two species include vervets, marmosets, and spider monkeys, as well as one ape, the orangutan. Another eighteen species, including our close relative the gorilla, exhibit slight signs. The remaining eighteen species, including baboons and our close relatives the chimpanzees, advertise ovulation conspicuously.

Step 2. Next, Sillén-Tullberg and Møller categorized the same sixty-eight species according to their mating system. Eleven species—including marmosets, gibbons, and many human societies—turn out to be monogamous. Twenty-three species—including other human societies, plus gorillas—have harems of females controlled by a single adult male. But the largest number of primate species—thirty-four, including vervets, bonobos, and chimpanzees—have a promiscuous system in which females routinely associate and copulate with multiple males.

Again I hear cries of Aha!—Why aren't humans also classified as promiscuous? Because I was careful to specify *routinely*. Yes, most woman have multiple sex partners in sequence over their lifetimes, and many women are at times involved with multiple men simultaneously. However, within any given estrus cycle the norm is for a woman to be involved with a single man, but the norm for a female vervet or bonobo is to be involved with several partners.

Step 3. As the next-to-last step, Sillén-Tullberg and Møller combined steps 1 and 2 to ask: is there any tendency for more or less conspicuous ovulations to be associated with a particular mating system? Based on a naive reading of our two competing theories, concealed ovulation should be a feature of monogamous species if the daddy-at-home theory is correct, but of promiscuous species if the many-fathers theory holds. In fact, the overwhelming majority of monogamous primate species analyzed—ten out of eleven species—prove to have concealed ovulation. Not a single monogamous primate species has boldly advertised ovulations, which instead are usually (in fourteen out of eighteen cases) confined to promiscuous species. That seems to be strong support for the daddy-at-home theory.

However, the fit between predictions and theory is only a half-fit, because the reverse correlations don't hold up at all. While most monogamous species have concealed ovulation, concealed ovulation in turn is no guarantee of monogamy. Out of thirty-two species with concealed ovulation, twenty-two aren't monogamous but are instead promiscuous or live in harems. Concealed ovulators include monogamous night monkeys, often-monogamous humans, harem-holding langur monkeys, and promiscuous vervets. Thus, whatever caused concealed ovulation to evolve in the first place, it can be maintained thereafter under the most varied mating systems.

Similarly, while most species with boldly advertised ovulations are promiscuous, promiscuity is no guarantee of advertisement. In fact, most promiscuous primates— twenty out of thirty-four species—either have concealed ovulation or only slight signs. Harem-holding species as well have invisible, slightly visible, or conspicuous ovulations, depending on the particular species. These complexities warn us that concealed ovulation will prove to serve different functions, according to the particular mating system with which it coexists.

Step 4. To identify these changes of function, Sillén-Tullberg and Møller got the bright idea of studying the family tree of living primate species. They thereby hoped to identify the points in primate evolutionary history at which there had been evolutionary changes in ovulatory signals and mating systems. The underlying rationale is that some modern species that are very closely related to each other, hence presumably derived recently from a common ancestor, turn out to differ in mating system or in strength of ovulatory signals. This implies recent evolutionary changes in mating systems or signals.

Here's an example of how the reasoning works. We know that humans, chimps, and gorillas are genetically about 98 percent identical and stem from an ancestor ("the Missing Link") that lived as recently as nine million years ago. Yet those three modern descendants of the Missing Link now exhibit all three types of ovulatory signal: concealed ovulation in humans, slight signals in gorillas, bold advertisement in chimps. Hence only one of those descendants can be like the Missing Link in ovulatory signals, and the other two descendants must have evolved different signals.

In fact, most living species of primitive primates have slight signs of ovulation. Hence the Missing Link may have preserved that condition, and gorillas may have inherited it in turn from the Missing Link (see figure 4.1). Within the last nine million years, though, humans must have evolved concealed ovulation, and chimps must have evolved bold advertisement. Our signals and those of chimps thus diverged in opposite directions from the cues of our mildly signaling ancestors. To us humans, the swollen derrières of ovulating chimps look like those of baboons. However, the

Family Tree of Ovulatory Signs

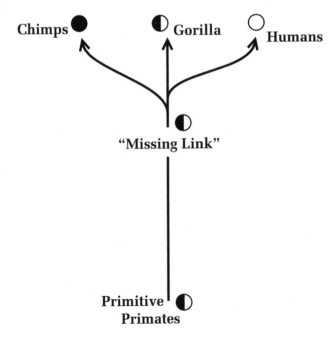

Figure 4.1

ancestors of chimps and baboons must have evolved their eye-catching derrières quite independently, since the ancestors of baboons and of the Missing Link parted company around thirty million years ago.

By similar reasoning, one can infer other points in the primate family tree at which ovulatory signals must have changed. It turns out that switches of signals have evolved at least twenty times. There have been at least three independent origins of bold advertisement (including the example in chimps); at least eight independent origins of concealed ovulation (including its origins in us, in orangutans, and in at least six separate groups of monkeys); and several *re*appearances of slight signs of ovulation, from either concealed ovulation (as in some howler monkeys) or from bold advertisement (as in many macaques).

In the same way as we've just seen for ovulatory signals, one can also identify points in the primate family tree at which mating systems must have changed. The original system for the common ancestor of all monkeys and apes was probably promiscuous mating. But if we now look at humans and our closest relatives, the chimps and gorillas, we find all three types of mating system represented: harems in gorillas, promiscuity in chimps, and either monogamy or harems in humans (see figure 4.2). Thus, among the three descendants of the Missing Link of nine million years ago, at least two must have changed their mating system. Other evidence suggests that the Missing Link lived in harems, so that gorillas and some human societies may just have retained that mating system. But chimps must have reinvented promiscuity, while many human societies invented monogamy. Again, we see that humans and chimps have evolved oppositely, in mating systems as in ovulatory signals.

Overall, it appears that monogamy has evolved independently at least seven times in higher primates: in us, in gibbons, and in at least five separate groups of monkeys.

Family Tree of Mating Systems

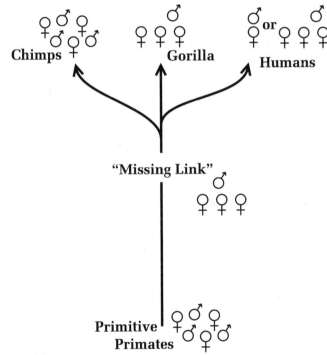

Figure 4.2

......

Harems must have evolved at least eight times, including in the Missing Link. Chimps and at least two monkeys must have reinvented promiscuity after their recent ancestors had given it up for harems.

• • • • •

Thus, we have reconstructed both the type of mating system and the type of ovulatory signal that probably existed in primates of the remote past, all along the primate family tree. We can now, finally, put both types of information together and ask: what mating system prevailed at each point in our family tree when concealed ovulation evolved?

Here's what one learns. Consider those ancestral species that gave signals of ovulation, and that then went on to lose those signals and evolve concealed ovulation. Only one of those ancestral species was monogamous. In contrast, eight, perhaps as many as eleven, of them were promiscuous or harem-holding species—one of them being the human ancestor that arose from the harem-holding Missing Link. We thus conclude that promiscuity or harems, not monogamy, is the mating system that leads to concealed ovulation (see figure 4.3). This is the conclusion predicted by the many-fathers theory. It doesn't agree with the daddy-at-home theory.

Conversely, we can also ask: what were the ovulatory signals prevailing at each point in our family tree when monogamy evolved? We find that monogamy never evolved in species with bold advertisement of ovulation. Instead, monogamy has usually arisen in species that already had concealed ovulation, and sometimes in species that already had slight ovulatory signals (see figure 4.4). This conclusion agrees with the predictions of the daddy-at-home theory.

How can these two apparently opposite conclusions be reconciled? Recall that Sillén-Tullberg and Møller found, in step 3 of their analysis, that almost all monogamous primates have concealed ovulation. We now see that that result must have arisen in two steps. *First,* concealed ovu-

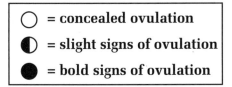

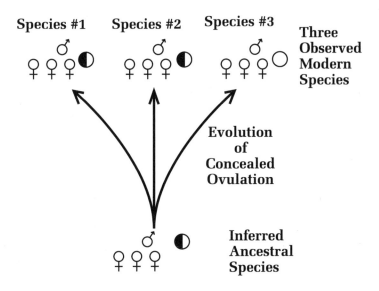

Figure 4.3
By combining facts about modern observed species with inferences about ancestral species, one can infer the mating system prevailing when ovulatory signals underwent evolutionary change. We infer that species 3 evolved concealed ovulation from a harem-holding ancestor with slight signs of ovulation, while species 1 and 2 preserved the ancestral mating system (harems) and slight ovulatory signs.

••••••

lation arose, in a promiscuous or harem-holding species. *Then,* with concealed ovulation already present, the species switched to monogamy (see figure 4.4).

The Evolution of Concealed Ovulation

mating system	harem	➔	harem	➔	monogamy
ovulatory signals	slight	➔	concealed	➔	concealed
function of ovulatory signals, or lack thereof	efficient sex		confuse paternity and prevent infanticide		keep daddy home

Figure 4.4

......

.....

Perhaps by now you're finding our sexual history confusing. We started out with an apparently simple question that deserved a simple answer: why do we hide our ovulations and have recreational sex on any day of the month? Instead of a simple answer, you're being told that the answer is more complex and involves two steps.

What it boils down to is that concealed ovulation has repeatedly changed, and actually reversed, its function during primate evolutionary history. It arose at a time when our ancestors were still promiscuous or living in harems. At such times, concealed ovulation let the ancestral ape-woman distribute her sexual favors to many males, none of which could swear that he was the father of her child but each of which knew that he might be. As a result, none of those potentially murderous males wanted to harm the ape-woman's baby, and some may actually have protected or helped feed it. Once the ape-woman had evolved concealed ovulation for that purpose, she then used it to pick a good caveman, to entice or force him to stay at home with her, and to get him to provide lots of protection or help for her baby—secure in the knowledge that it was his baby too.

On reflection, we shouldn't be surprised at this shift of function for concealed ovulation. Such shifts are very common in evolutionary biology. That's because natural selection doesn't proceed consciously and in a straight line toward a distant perceived goal, in the way that an engineer consciously designs a new product. Instead, a feature that serves one function in an animal begins to serve another function as well, becomes modified as a result, and may even lose the original function. The consequence is frequent reinventions of similar adaptations, and frequent losses, shifts, or even reversals of function, as living things evolve.

One of the most familiar examples involves vertebrate limbs. The fins of ancestral fishes, used for swimming, evolved into the legs of ancestral reptiles, birds, and mammals, which used them for running or hopping on land. The front legs of certain ancestral mammals and reptile-birds subsequently evolved into the wings, used for flying, of bats and modern birds, respectively. Bird wings and mammal legs then evolved independently into the flippers of penguins and whales, respectively, thereby reverting to a swimming function and effectively reinventing the fins of fish. At least three groups of fish descendants independently lost their limbs to become snakes, legless lizards, and the legless amphibians known as cecilians. In essentially the same way, features of reproductive biology—such as concealed ovulation, boldly advertised ovulation, monogamy, harems, and promiscuity—have repeatedly changed function and been transmuted into each other, reinvented, or lost.

The implications of these evolutionary shifts can lend

zest to our love lives. For example, in the last novel by the great German writer Thomas Mann, *Confessions of Felix Krull, Confidence Man,* Felix shares a compartment on a train journey with a paleontologist, who regales him with an account of vertebrate limb evolution. Felix, an accomplished and imaginative ladies' man, is delighted by the implications. "Human arms and legs retain the bones of the most primitive land animals! . . . It's thrilling! . . . A woman's shapely charming arm, which embraces us if we find favor . . . it's no different from the primordial bird's clawed wing, and the fish's pectoral fin. . . . I'll think of that, next time. . . . Dream of that shapely arm, with its ancient scaffolding of bones!"

Now that Sillén-Tullberg and Møller have unraveled the evolution of concealed ovulation, you can nourish your own fantasy with its implications, just as Felix Krull nourished his fantasy with the implications of vertebrate limb evolution. Wait until the next time that you are having sex for fun, at a nonfertile time of the ovulatory cycle, while enjoying the security of a lasting monogamous relationship. At such a time, reflect on how your bliss is made paradoxically possible by precisely those features of your physiology that distinguished your remote ancestors as they languished in harems, or as they rotated among promiscuously shared sex partners. Ironically, those wretched ancestors had sex only on rare days of ovulation, when they perfunctorily discharged the biological imperative to fertilize, robbed of your leisurely pleasure by their desperate need for swift results.

CHAPTER 5

WHAT ARE MEN GOOD FOR?

The Evolution of Men's Roles

Last year I received a remarkable letter from a professor at a university in a distant city, inviting me to an academic conference. I did not know the writer, and I couldn't even figure out from the name whether the writer was a man or a woman. The conference would involve long plane flights and a week away from home. However, the letter of invitation was beautifully written. If the conference was going to be as beautifully organized, it might be exceptionally interesting. With some ambivalence because of the time commitment, I accepted.

My ambivalence vanished when I arrived at the conference, which turned out to be every bit as interesting as I had anticipated. In addition, much effort had been made to arrange outside activities for me, including shopping, bird-watching, banquets, and tours of archaeological sites. The professor behind this masterpiece of organization and the original virtuoso letter proved to be a woman. In addition to giving a brilliant lecture at the conference and being a very pleasant person, she was among the most stunningly beautiful women I had ever met.

On one of the shopping trips that my hostess arranged, I bought several presents for my wife. The student who had been sent along as my guide evidently reported these purchases to my hostess, because she commented on them when I sat next to her at the conference banquet. To my astonishment, she told me, "My husband never buys me any presents!" She had formerly bought presents for him but eventually stopped when he never reciprocated.

Someone across the table then asked me about my fieldwork on birds of paradise in New Guinea. I explained that male birds of paradise provide no help in rearing the nestlings but instead devote their time to trying to seduce as many females as possible. Surprising me again, my hostess burst out, "Just like men!" She explained that her own husband was much better than most men, because he encouraged her career aspirations. However, he spent most evenings with other men from his office, watched television while at home on the weekend, and avoided helping with the household and with their two children. She had repeatedly asked him to help; she finally gave up and hired a housekeeper. There is, of course, nothing unusual about this story. It stands out in my mind only because this woman was so beautiful, nice, and talented that one might naively have expected the man who chose to marry her to have remained interested in spending time with her.

My hostess nevertheless enjoys much better domestic conditions than do many other wives. When I first began to work in the New Guinea highlands, I often felt enraged at the sight of gross abuse of women. Married couples whom I encountered along jungle trails typically consisted of a woman bent under an enormous load of firewood, vegetables, and an infant, while her husband sauntered along upright, bearing nothing more than his bow and arrow. Men's hunting trips seemed to yield little more than male bonding opportunities, plus some prey animals immediately consumed in the jungle by the men. Wives were bought, sold, and discarded without their consent.

Later, though, when I had children of my own and sensed my feelings as I shepherded my family on walks, I thought that I could better understand the New Guinea men striding beside their families. I found myself striding next to my own children, devoting all my attention to making sure that they did not get run over, fall, wander off, or suffer some other mishap. Traditional New Guinea men had to be even more attentive because of the greater risks facing their children and wives. Those seemingly carefree men strolling along beside a heavily burdened wife were actually functioning as lookouts and protectors, keeping their hands free so that they could quickly deploy their bow and arrow in the event of ambush by men of another tribe. But the men's hunting trips, and the sale of women as wives, continue to trouble me.

To ask what men are good for may sound like a flip one-liner. In fact, the question touches a raw nerve in our society. Women are becoming intolerant of men's self-ascribed status and are criticizing those men who provide better for themselves than for their wives and children. The question also poses a big theoretical problem for anthropologists. By the criterion of services offered to mates and children, males of most mammal species are good for nothing except injecting sperm. They part from the female after copulation, leaving her to bear the entire burden of feeding, protecting, and training the offspring. But human males differ by (usually or often) remaining with their mate and offspring after copulation. Anthropologists widely assume that men's resulting added roles contributed crucially to the evolution of our species' most distinctive features. The reasoning goes as follows.

The economic roles of men and women are differentiated in all surviving hunter-gatherer societies, a category that encompassed all human societies until the rise of agriculture ten thousand years ago. Men invariably spend more time hunting large animals, while women spend more time gathering plant foods and small animals and caring for children. Anthropologists traditionally view this ubiquitous differentiation as a division of labor that promotes the nuclear family's joint interests and thereby represents a sound strategy of cooperation. Men are much better able than women to track and kill big animals, for the obvious reasons that men don't have to carry infants around to nurse them and that men are on the average more muscular than women. In the view of anthropologists, men hunt in order to provide meat to their wives and children.

A similar division of labor persists in modern industrial societies: many women still devote more time to child care than men do. While men no longer hunt as their main occupation, they still bring food to their spouse and children by holding money-paying jobs (as do a majority of American women as well). Thus, the expression "bringing home the bacon" has a profound and ancient meaning.

Meat provisioning by traditional hunters is considered a distinctive function of human males, shared with only a few of our fellow mammal species such as wolves and African hunting dogs. It is commonly assumed to be linked to other universal features of human societies that distinguish us from our fellow mammals. In particular, it is linked to the fact that men and women remain associated in nuclear families after copulation, and that human children (unlike young apes) remain unable to obtain their own food for many years after weaning.

This theory, which seems so obvious that its correctness is generally taken for granted, makes two straightforward predictions about men's hunting. First, if the main purpose of hunting is to bring meat to the hunter's family, men should pursue the hunting strategy that reliably yields the most meat. Hence we should observe that men are on the average bagging more pounds of meat per day by going after big animals than they would bring home by targeting small animals. Second, we should observe that a hunter brings his kill to his wife and kids, or at least shares it preferentially with them rather than with nonrelatives. Are these two predictions true?

.....

Surprisingly for such basic assumptions of anthropology, these predictions have been little tested. Perhaps unsurprisingly, the lead in testing them has been taken by a woman anthropologist, Kristen Hawkes of the University of Utah. Hawkes's tests have been based especially on quantitative measurements of foraging yields for Paraguay's Northern Aché Indians, carried out jointly with Kim Hill, A. Magdalena Hurtado, and H. Kaplan. Hawkes performed other tests on Tanzania's Hadza people in collaboration with Nicholas Blurton Jones and James O'Connell. Let's consider first the evidence for the Aché.

The Northern Aché used to be full-time hunter-gatherers and continued to spend much time foraging in the forest even after they began to settle at mission agricultural settlements in the 1970s. In accord with the usual human pattern, Aché men specialize in hunting large mammals, such as peccaries and deer, and they also collect masses of honey from bees' nests. Women pound starch from palm trees, gather fruits and insect larvae, and care for children. An Aché man's hunting bag varies greatly from day to day: he brings home food enough for many people if he kills a peccary or finds a beehive, but he gets nothing at all on one-quarter of the days he spends hunting. In contrast, women's returns are predictable and vary little from day to day because palms are abundant; how much starch a woman gets is mainly a function of just how much time she spends pounding it. A woman can always count on getting enough for herself and her children, but she can never reap a bonanza big enough to feed many others.

The first surprising result from the studies by Hawkes and her colleagues concerned the difference between the returns achieved by men's and women's strategies. Peak yields were, of course, much higher for men than for women, since a man's daily bag topped 40,000 calories when he was lucky enough to kill a peccary. However, a man's average daily return of 9,634 calories proved to be lower than that of a woman (10,356), and a man's median return (4,663 calories per day) was much lower. The reason for this paradoxical result is that the glorious days when a man bagged a peccary were greatly outnumbered by the humiliating days when he returned empty-handed.

Thus, Aché men would do better in the long run by sticking to the unheroic "woman's job" of pounding palms than by their devotion to the excitement of the chase. Since men are stronger than women, they could pound even more daily calories of palm starch than can women, if they chose to do so. In going for high but very unpredictable stakes, Aché men can be compared to gamblers who aim for the jackpot: in the long run, gamblers would do much better by putting their money in the bank and collecting the boringly predictable interest.

The other surprise was that successful Aché hunters do not bring meat home mainly for their wives and kids but share it widely with anyone around. The same is true for men's finds of honey. As a result of this widespread sharing, three-quarters of all the food that an Aché consumes is acquired by someone outside his or her nuclear family.

It's easy to understand why Aché women aren't big-game hunters: they can't spend the time away from their children, and they can't afford the risk of going even a day with an empty bag, which would jeopardize lactation and pregnancy. But why does a man eschew palm starch, settle for the lower average return from hunting, and not bring home his catch to his wife and kids, as the traditional view of anthropologists predicts?

This paradox suggests that something other than the best interests of his wife and children lie behind an Aché man's preference for big-game hunting. As Kristen Hawkes described these paradoxes to me, I developed an awful foreboding that the true explanation might prove less noble than the male's mystique of bringing home the bacon. I began to feel defensive on behalf of my fellow men and to search for explanations that might restore my faith in the nobility of the male strategy.

My first objection was that Kristen Hawkes's calculations of hunting returns were measured in calories. In reality, any nutritionally aware modern reader knows that not all calories are equal. Perhaps the purpose of big-game hunting lies in fulfilling our need for protein, which is more valuable to us nutritionally than the humble carbohydrates of palm starch. However, Aché men target not only protein-rich meat but also honey, whose carbohydrates are every bit as humble as those of palm starch. While Kalahari San men ("Bushmen") are hunting big game, San women are gathering and preparing mongongo nuts, an excellent protein source. While lowland New Guinea hunter-gatherer men are wasting their days in the usually futile search for kangaroos, their wives and children are predictably acquiring protein in the form of fish, rats, grubs, and spiders. Why don't San and New Guinea men emulate their wives?

I next began to wonder whether Aché men might be unusually ineffective hunters, an aberration among modern hunter-gatherers. Undoubtedly, the hunting skills of Inuit (Eskimo) and Arctic Indian men are indispensable, especially in winter, when little food other than big game is available. Tanzania's Hadza men, unlike the Aché, achieve higher average returns by hunting big game rather than small game. But New Guinea men, like the Aché, persist in hunting even though yields are very low. And Hadza hunters persist in the face of enormous risks, since on the average they bag nothing at all on twenty-eight out of twenty-nine days spent hunting. A Hadza family could starve while waiting for the husband-father to win his gamble of bringing down a giraffe. In any case, all that meat occasionally bagged by a Hadza or Aché hunter isn't reserved for his family, so the question of whether big-game hunting yields higher or lower returns than alternative strategies is academic from his family's point of view. Big-game hunting just isn't the best way to feed a family.

Still seeking to defend my fellow men, I then wondered: could the purpose of widely sharing meat and honey be to smooth out hunting yields by means of reciprocal altruism? That is, I expect to kill a giraffe only every twenty-ninth day, and so does each of my hunter friends, but we all go off in different directions, and each of us is likely to kill his giraffe on a different day. If successful hunters agree to share meat with each other and their families, all of them will often have full bellies. By that interpretation, hunters should prefer to share their catch with the best other hunters, from whom they are most likely to receive meat some other day in return.

In reality, though, successful Aché and Hadza hunters share their catch with anyone around, whether he's a good or hopeless hunter. That raises the question of why an Aché or Hadza man bothers to hunt at all, since he can claim a share of meat even if he never bags anything himself. Conversely, why should he hunt when any animal that he kills will be shared widely? Why doesn't he just gather nuts and rats, which he can bring to his family and would not have to share with anyone else? There must be some ignoble motive for male hunting that I was overlooking in my efforts to find a noble motive.

As another possible noble motive, I thought that widespread sharing of meat helps the hunter's whole tribe, which is likely to flourish or perish together. It's not enough to concentrate on nourishing your own family if the rest of your tribe is starving and can't fend off an attack by tribal enemies. This possible motive, though, returns us to the original paradox: the best way for the whole Aché tribe to become well nourished is for everybody to humble themselves by pounding good old reliable palm starch and collecting fruit or insect larvae. The men shouldn't waste their time gambling on the occasional peccary.

In a last effort to detect family values in men's hunting, I reflected on hunting's relevance to the role of men as protectors. The males of many territorial animal species, such as songbirds, lions, and chimpanzees, spend much time patrolling their territories. Such patrols serve multiple purposes: to detect and expel intruding rival males from adjacent territories; to observe whether adjacent territories are in turn ripe for intrusion; to detect predators that could endanger the male's mate and offspring; and to monitor seasonal changes in abundance of foods and other resources. Similarly, at the same time as human hunters are looking for game, they too are attentive to potential dangers and opportunities for the rest of the tribe. In addition, hunting provides a chance to practice the fighting skills that men employ in defending their tribe against enemies.

This role of hunting is undoubtedly an important one. Nevertheless, one has to ask what specific dangers the hunters are trying to detect, and whose interests they are thereby trying to advance. While lions and other big carnivores do pose dangers to people in some parts of the world, by far the greatest danger to traditional hunter-gatherer human societies everywhere has been posed by hunters from rival tribes. Men of such societies were involved in intermittent wars, the purpose of which was to kill men of other tribes. Captured women and children of defeated rival tribes were either killed or else spared and acquired as wives and slaves, respectively. At worst, patrolling groups of male hunters could thus be viewed as advancing their own genetic self-interest at the expense of rival groups of men. At best, they could be viewed as protecting their wives and children, but mainly against the dangers posed by other men. Even in the latter case, the harm and the good that adult men bring to the rest of society by their patrolling activities would be nearly equally balanced.

•••••

Thus, all five of my efforts to rescue Aché big-game hunting as a sensible way for men to contribute nobly to the best interests of their wives and children collapsed. Kristen Hawkes then reminded me of some painful truths about how an Aché man himself (as opposed to his wife and kids) gets big benefits from his kills besides the food entering his stomach.

To begin with, among the Aché, as among other peoples, extramarital sex is not uncommon. Dozens of Aché women, asked to name the potential fathers (their sex partners around the time of conception) of 66 of their children, named an average of 2.1 men per child. Among a sample of 28 Aché men, women named good hunters more often than poor hunters as their lovers, and they named good hunters as potential fathers of more children.

To understand the biological significance of adultery, recall that the facts of reproductive biology discussed in chapter 2 introduce a fundamental asymmetry into the interests of men and women. Having multiple sex partners contributes nothing directly to a woman's reproductive output. Once a woman has been fertilized by one man, having sex with another man cannot lead to another baby for at least nine months, and probably for at least several years under hunter-gatherer conditions of extended lactational amenorrhea. In just a few minutes of adultery, though, an otherwise faithful man can double the number of his own offspring.

Now compare the reproductive outputs of men pursuing the two different hunting strategies that Hawkes terms the "provider" strategy and the "show-off" strategy. The provider hunts for foods yielding moderately high returns with high predictability, such as palm starch and rats. The show-off hunts for big animals; by scoring only occasional bonanzas amid many more days of empty bags, his mean return is lower. The provider brings home on the average the most food for his wife and kids, although he never acquires enough of a surplus to feed anyone else. The show-off on the average brings less food to his wife and kids but does occasionally have lots of meat to share with others.

Obviously, if a woman gauges her genetic interests by the number of children whom she can rear to maturity, that's a function of how much food she can provide them, so she is best off marrying a provider. But she is further well served by having show-offs as neighbors, with whom she can trade occasional adulterous sex for extra meat supplies for herself and her kids. The whole tribe also likes a show-off because of the occasional bonanzas that he brings home for sharing.

As for how a man can best advance his own genetic interests, the show-off enjoys advantages as well as disadvantages. One advantage is the extra kids he sires adulterously. The show-off also gains some advantages apart from adultery, such as prestige in his tribe's eyes. Others in the tribe want him as a neighbor because of his gifts of meat, and they may reward him with their daughters as mates. For the same reason, the tribe is likely to give favored treatment to the show-off's children. Among the disadvantages to the show-off are that he brings home on the average less food to his own wife and kids; this means that fewer of his legitimate children may survive to maturity. His wife may also philander while he is doing so, with the result that a lower percentage of her children are actually his. Is the show-off better off giving up the provider's certainty of paternity of a few kids, in return for the possibility of paternity of many kids?

The answer depends on several numbers, such as how many extra legitimate kids a provider's wife can rear, the percentage of a provider's wife's kids that are illegitimate, and how much a show-off's kids find their chances of survival increased by their favored status. The values of these numbers must differ among tribes, depending on the local ecology. When Hawkes estimated the values for the Aché, she concluded that, over a wide range of likely conditions, show-offs can expect to pass on their genes to more surviving children than can providers. This purpose, rather than the traditionally accepted purpose of bringing home the bacon to wife and kids, may be the real reason behind big-game hunting. Aché men thereby do good for themselves rather than for their families.

Thus, it is not the case that men hunters and women gatherers constitute a division of labor whereby the nuclear family as a unit most effectively promotes its joint interests, and whereby the work force is selectively deployed for the good of the group. Instead, the hunter-gatherer lifestyle involves a classic conflict of interest. As I discussed in chapter 2, what's best for a man's genetic interests isn't necessarily best for a woman's, and vice versa. Spouses share interests, but they also have divergent interests. A woman is best off married to a provider, but a man is not best off being a provider.

Biological studies of recent decades have demonstrated numerous such conflicts of interest in animals and humans—not only conflicts between husbands and wives (or between mated animals), but also between parents and children, between a pregnant woman and her fetus, and between siblings. Parents share genes with their offspring, and siblings share genes with each other. However, siblings are also potentially each other's closest competitors, and parents and offspring also potentially compete. Many animal studies have shown that rearing offspring reduces the parent's life expectancy because of the energy drain and risks that the parent incurs. To a parent, an offspring represents one opportunity to pass on genes, but the parent may have other such opportunities. The parent's interests may be better served by abandoning one offspring and devoting resources to other offspring, whereas the offspring's interests may be best served by surviving at the expense of its parents. In the animal world as in the human world, such conflicts not infrequently lead to infanticide, parricide (the murder of parents by an offspring), and siblicide (the murder of one sibling by another). While biologists explain the conflicts by theoretical calculations based on genetics and foraging ecology, all of us recognize them from experience, without doing any calculations. Conflicts of interest between people closely related by blood or marriage are the commonest, most gut-wrenching tragedies of our lives.

. . . . .

What general validity do these conclusions possess? Hawkes and her colleagues studied just two hunter-gatherer peoples, the Aché and the Hadza. The resulting conclusions await testing of other hunter-gatherers. The answers are likely to vary among tribes and even among individuals. From my own experience in New Guinea, Hawkes's conclusions are likely to apply even more strongly there. New Guinea has few large animals, hunting yields are low, and bags are often empty. Much of the catch is consumed directly by the men while off in the jungle, and the meat of any big animal brought home is shared widely. New Guinea hunting is hard to defend economically, but it brings obvious payoffs in status to successful hunters.

What about the relevance of Hawkes's conclusions to our own society? Perhaps you're already livid because you foresaw that I'd raise that question, and you're expecting me to conclude that American men aren't good for much. Of course that's not what I conclude. I acknowledge that many (most? by far the most?) American men are devoted husbands, work hard to increase their income, devote that income to their wives and kids, do much child care, and don't philander.

But, alas, the Aché findings are relevant to at least some men in our society. Some American men do desert their wives and children. The proportion of divorced men who renege on their legally stipulated child support is scandalously high, so high that even our government is starting to do something about it. Single parents outnumber coparents in the United States, and most single parents are women.

Among those men who remain married, all of us know some who take better care of themselves than of their wives and children, and who devote inordinate time, money, and energy to philandering and to male status symbols and activities. Typical of such male preoccupations are cars, sports, and alcohol consumption. Much bacon isn't brought home. I don't claim to have measured what percentage of American men rate as show-offs rather than providers, but the percentage of show-offs appears not to be negligible.

Even among devoted working couples, time budget studies show that American working women spend on the average twice as many hours on their responsibilities (defined as job plus children plus household) as do their husbands, yet women receive on the average less pay for the same job. When American husbands are asked to estimate the number of hours that they and their wives each devote to children and household, the same time budget studies show that men tend to overestimate their own hours and to underestimate their wife's hours. It's my impression that men's household and child-care contributions are on the average even lower in some other industrialized countries, such as Australia, Japan, Korea, Germany, France, and Poland, to mention just a few with which I happen to be familiar. That's why the question what men are good for continues to be debated within our societies, as well as between anthropologists.

....

CHAPTER 6

MAKING MORE BY MAKING LESS

The Evolution of Female Menopause

Most wild animals remain fertile until they die, or until close to that time. So do human males: although some men become infertile or less fertile at various ages for various reasons, men experience no universal shutdown of fertility at any particular age. There are innumerable well-attested cases of old men, including a ninety-four-year-old, fathering children.

But human females undergo a steep decline in fertility from around age forty, leading to universal complete sterility within a decade or so. While some women continue to have regular menstrual cycles up to the age of fifty-four or fifty-five, conception after the age of fifty was rare until the recent development of medical technologies using hormone therapy and artificial fertilization. For example, among the American Hutterites, a strict religious community that is well nourished and opposed to contraception, women produce babies as fast as is biologically possible for humans, with a mean interval of only two years between births, and a mean final number of eleven children. Even Hutterite women stop producing babies by age forty-nine.

To laypeople, menopause is an inevitable fact of life, albeit often a painful one anticipated with foreboding. But to evolutionary biologists, human female menopause is an aberration in the animal world and an intellectual paradox. The essence of natural selection is that it promotes genes for traits that increase the number of one's descendants bearing those genes. How could natural selection possibly result in every female member of a species carrying genes that throttle her ability to leave more descendants? All biological traits are subject to genetic variation, including the age of human female menopause. Once female menopause somehow became fixed in humans for whatever reason, why did not its age of onset gradually become pushed back until it disappeared again, because those women who experienced menopause later in life left behind more descendants?

To evolutionary biologists, female menopause is thus among the most bizarre features of human sexuality. As I shall argue, it is also among the most important. Along with our big brains and upright posture (emphasized in every text of human evolution), and our concealed ovulations and penchant for recreational sex (to which texts devote less attention), I believe that female menopause was among the biological traits essential for making us distinctively human—a creature more than, and qualitatively different from, an ape.

.....

Many biologists would balk at what I have just said. They would argue that human female menopause does not pose an unsolved problem, and that there is no need to discuss it further. Their objections are of three types.

First, some biologists dismiss human female menopause as an artifact of a recent increase in human expected life span. That increase stems not just from public health measures within the last century but possibly also from the rise of agriculture ten thousand years ago, and even more likely from evolutionary changes leading to increased human survival skills within the last forty thousand years. According to this view, menopause could not have been a frequent occurrence for most of the several million years of human evolution, because (supposedly) almost no women or men survived past the age of forty. Of course, the female reproductive tract was programmed to shut down by age forty, because it would not have had the opportunity to operate thereafter anyway. The increase in human life span has developed much too recently in our evolutionary history for the female reproductive tract to have had time to adjust—so goes this objection.

However, this view ignores the fact that the human male reproductive tract, and every other biological function of both women and men, continue to function in most people for many decades after age forty. One would therefore have to assume that every other biological function was able to adjust quickly to our new long life span, leaving unexplained why female reproduction was uniquely incapable of doing so. The claim that formerly few women survived until the age of menopause is based on paleodemography, that is, on attempts to estimate age at time of death in ancient skeletons. Those estimates rest on unproven, implausible assumptions, such as that the recovered skeletons represent an unbiased sample of an entire ancient population, or that ancient adult skeletons really can be aged accurately. While paleodemographers' ability to distinguish the ancient skeleton of a ten-year-old from that of a twenty-five-year-old is not in question, the ability they claim to distinguish an ancient forty-year-old from a fifty-five-year-old has never been demonstrated. One can hardly reason by comparison with skeletons of modern people, whose different lifestyles, diets, and diseases surely make their bones age at different rates from the bones of ancients.

A second objection acknowledges human female menopause as a possibly ancient phenomenon but denies that it is unique to humans. Many or most wild animals exhibit a decrease in fertility with age. Some elderly individuals of a wide variety of wild mammal and bird species are found to be infertile. Many elderly female individuals of rhesus macaques and certain strains of laboratory mice, living in laboratory cages or zoos where their lives are considerably extended over expected spans in the wild by gourmet diets, superb medical care, and complete protection from enemies, do become infertile. Hence some biologists object that human female menopause is merely part of a widespread phenomenon of animal menopause. Whatever that phenomenon's explanation, its existence in many species would mean that there is not necessarily anything peculiar about menopause in the human species requiring explanation.

However, one swallow does not make a summer, nor does one sterile female constitute menopause. That is, detection of an occasional sterile elderly individual in the wild, or of regular sterility in caged animals with artificially extended life spans, does nothing to establish the existence of menopause as a biologically significant phenomenon in the wild. That would require demonstrating that a substantial fraction of adult females in a wild animal population become sterile and spend a significant portion of their life spans after the end of their fertility.

The human species does fulfill that definition, but only one or possibly two wild animal species are definitely known to do so. One is an Australian marsupial mouse in which *males* (not females) exhibit something like menopause: all males in the population become sterile within a short time in August and die over the next couple of weeks, leaving a population that consists solely of pregnant females. In that case, however, the postmenopausal phase is a negligible fraction of the total male life span. Marsupial mice do not exemplify true menopause but are more appropriately considered an example of big-bang reproduction, alias semelparity—a single lifetime reproductive effort rapidly followed by sterility and death, as in salmon and century plants. The better example of animal menopause is provided by pilot whales, among which one-quarter of all adult females killed by whalers proved to be postmenopausal, as judged by the condition of their ovaries. Female pilot whales enter menopause at the age of thirty or forty years, have a mean survival of at least fourteen years after menopause, and may live for over sixty years.

Menopause as a biologically significant phenomenon is thus not unique to humans, being shared at least with one species of whale. It would be worth looking for evidence of menopause in killer whales and a few other species as possible candidates. But still-fertile elderly females are often encountered among well-studied wild populations of other long-lived mammals, including chimpanzees, gorillas, baboons, and elephants. Hence those species and most others are unlikely to be characterized by regular menopause. For example, a fifty-five-year-old elephant is considered elderly, since 95 percent of elephants die before that age. But the fertility of fifty-five-year-old female elephants is still half that of younger females in their prime.

Thus, female menopause is sufficiently unusual in the animal world that its evolution in humans requires explanation. We certainly did not inherit it from pilot whales, from whose ancestors our own ancestors parted company over fifty million years ago. In fact, we must have evolved it since our ancestors separated from those of chimps and gorillas seven million years ago, because we undergo menopause and chimps and gorillas appear not to (or at least not regularly).

The third and last objection acknowledges human menopause as an ancient phenomenon that is unusual among animals. Instead, these critics say that we need not seek an explanation for menopause, because the puzzle has already been solved. The solution (they say) lies in the physiological mechanism of menopause: a woman's egg supply is fixed at her birth and not added to later in her life. One or more eggs are lost by ovulation at each menstrual cycle, and far more eggs simply die (termed atresia). By the time a woman is fifty years old, most of her original egg supply has been depleted. Those eggs that remain are half a century old, increasingly unresponsive to pituitary hormones, and too few in number to produce enough estradiol to trigger the release of pituitary hormones.

But there is a fatal counterobjection to this objection. While the objection is not wrong, it is incomplete. Yes, depletion and aging of the egg supply are the immediate causes of human menopause, but why did natural selection program women such that their eggs become depleted or unresponsive in their forties? There is no compelling reason why we could not have evolved twice as large a starting quota of eggs, or eggs that remain responsive after half a century. The eggs of elephants, baleen whales, and possibly albatrosses remain viable for at least sixty years, and the eggs of tortoises are viable for much longer, so human eggs could presumably have evolved the same capability.

The basic reason why the third objection is incomplete is because it confuses proximate mechanisms with ultimate causal explanations. (A proximate mechanism is an immediate direct cause, while an ultimate explanation is the last in the long chain of factors leading up to that immediate cause. For example, the proximate cause of a marriage breakup may be a husband's discovery of his wife's extramarital affairs, but the ultimate explanation may be the husband's chronic insensitivity and the couple's basic incompatibility that drove the wife to affairs.) Physiologists and molecular biologists regularly fall into the trap of overlooking this distinction, which is fundamental to biology, history, and human behavior. Physiology and molecular biology can do no more than identify proximate mechanisms; only evolutionary biology can provide ultimate causal explanations. As one simple example, the proximate reason why so-called poison-dart frogs are poisonous is that they secrete a lethal chemical named batrachotoxin. But that molecular biological mechanism for the frogs' poisonousness could be considered an unimportant detail because many other poisonous chemicals would have worked equally well. The ultimate causal explanation is that poison-dart frogs evolved poisonous chemicals because they are small, otherwise defenseless animals that would be easy prey for predators if they were not protected by poison.

We have already seen repeatedly in this book that the big questions about human sexuality are the evolutionary questions about ultimate causal explanation, not the search for proximate physiological mechanisms. Yes, sex is fun for us because women have concealed ovulations and are constantly receptive, but why did they evolve that unusual reproductive physiology? Yes, men have the physiological capacity to produce milk, but why did they not evolve to exploit that capacity? For menopause as well, the easy part of the puzzle is the mundane fact that a woman's egg supply gets depleted or impaired by around the time she is fifty years old. The challenge is to understand why we evolved that seemingly self-defeating detail of reproductive physiology.

• • • • •

The aging (or senescence, as biologists call it) of the female reproductive tract cannot be profitably considered in isolation from other aging processes. Our eyes, kidneys, heart, and all other organs and tissues also senesce. But that aging of our organs is not physiologically inevitable—or at least it's not inevitable that they senesce as rapidly as they do in the human species, because the organs of some turtles, clams, and other species remain in good condition much longer than ours do.

Physiologists and many other researchers on aging tend to search for a single all-encompassing explanation of aging. Popular explanations hypothesized in recent decades have invoked the immune system, free radicals, hormones, and cell division. In reality, though, all of us over forty know that everything about our bodies gradually deteriorates, and not just our immune systems and our defenses against free radicals. Although I have had a less stressful life and better medical care than most of the world's nearly six billion people, I can still tick off the aging processes that have already taken their toll on me by age fifty-nine: impaired hearing at high pitch, failure of my eyes to focus at short distances, less acute senses of smell and taste, loss of one kidney, tooth wear, less flexible fingers, and so on. My recovery from injuries is already slower than it used to be: I had to give up running because of recurrent calf injuries, I recently completed a slow recovery from a left elbow injury, and now I have just injured the tendon of a finger. Ahead of me, if the experience of other men is any guide, lies the familiar litany of complaints, including heart disorders, clogged arteries, bladder trouble, joint problems, prostate enlargement, memory loss, colon cancer, and so on. All that deterioration is what we mean by aging.

The basic reasons behind this grim litany are easily understood by analogy to human-built structures. Animal bodies, like machines, tend to deteriorate gradually or become acutely damaged with age and use. To combat those tendencies, we consciously maintain and repair our machines. Natural selection ensures that our body unconsciously maintains and repairs itself.

Both bodies and machines are maintained in two ways. First, we repair a part of a machine when it is acutely damaged. For example, we fix a car's punctured tire or bashed-in fender, and we replace its brakes or tires if they become damaged beyond repair. Our body similarly repairs acute damage. The most visible example is wound repair when we cut our skin, but molecular repair of damaged DNA and many other repair processes go on invisibly inside us. Just as a ruined tire can be replaced, our body has some capacity to regenerate parts of damaged organs such as by making new kidney, liver, and intestinal tissue. That capacity for regeneration is much better developed in many other animals. If only we were like starfish, crabs, sea cucumbers, and lizards, which can regenerate their arms, legs, intestines, and tail, respectively!

The other type of upkeep of machines and bodies is regular or automatic maintenance to reverse gradual wear, regardless of whether there has been any acute damage. For example, at times of scheduled maintenance we change our car's motor oil, spark plugs, fan belt, and ball bearings. Similarly, our body constantly grows new hair, replaces the lining of the small intestine every few days, replaces our red blood cells every few months, and replaces each tooth once in our lifetime. Invisible replacement goes on for the individual protein molecules that make up our bodies.

How well you maintain your car, and how much money or resources you put into its maintenance, strongly influence how long it lasts. The same can be said of our bodies, not only with respect to our exercise programs, visits to the doctor, and other conscious maintenance, but also with respect to the unconscious repair and maintenance that our bodies do on themselves. Synthesizing new skin, kidney tissue, and proteins uses up a lot of biosynthetic energy. Animal species vary greatly in their investment in self-maintenance, hence in the rate at which they senesce. Some turtles live for over a century. Laboratory mice, living in cages with abundant food and no predators or risks, and receiving better medical care than any wild turtle or the vast majority of the world's people, inevitably become decrepit and die of old age before their third birthday. There are aging differences even among us humans and our closest relatives, the great apes. Well-nourished apes living in the safety of zoo cages and attended by veterinarians rarely (if ever) live past age sixty, while white Americans exposed to much greater danger and receiving less medical attention now live to an average of seventy-eight years for men, eighty-three years for women. Why do our bodies unconsciously take better care of themselves than do apes' bodies? Why do turtles senesce so much more slowly than mice?

We could avoid aging entirely and (barring accidents) live forever if we went all out for repair and changed all the parts of our bodies frequently. We could avoid arthritis by growing new limbs, as crabs do, avoid heart attacks by periodically growing a new heart, and minimize tooth decay by regrowing new teeth five times (as elephants do, instead of just once, as we do). Some animals thus make a big investment in certain aspects of body repair, but no animal makes a big investment in all aspects, and no animal avoids aging entirely.

Analogy to our cars again makes the reason obvious: the expense of repair and maintenance. Most of us have only limited amounts of money, which we are obliged to budget. We put just enough money into car repair to keep our car running as long as it makes economic sense to do so. When the repair bills get too high, we find it cheaper to let the old car die and buy a new one. Our genes face a similar trade-off between repairing the old body that contains the genes and making new containers for the genes (that is, babies). Resources spent on repair, whether of cars or of bodies, eat away at the resources available for buying new cars or making babies. Animals with cheap self-repair and short life spans, like mice, can churn out babies much more rapidly than can expensive-to-maintain, long-lived animals like us. A female mouse that will die at the age of two, long before we humans achieve fertility, has been producing five babies every two months since she was a few months old.

That is, natural selection adjusts the relative investments in repair and reproduction so as to maximize the transmission of genes to offspring. The balance between repair and reproduction differs between species. Some species stint on repair and churn out babies quickly but die early, like mice. Other species, like us, invest heavily in repair, live for nearly a century, and can produce a dozen babies in that time (if you are a Hutterite woman), or over a thousand babies (if you are Emperor Moulay the Bloodthirsty). Your annual rate of baby production is lower than the mouse's (even if you are Moulay) but you have more years in which to do it.

·····

It turns out that an important evolutionary determinant of biological investment in repair—hence of life span under the best possible conditions—is the risk of death from accidents and bad conditions. You don't waste money maintaining your taxi if you are a taxi driver in Teheran, where even the most careful taxi driver is bound to suffer a major fender-bender every few weeks. Instead, you save your money to buy the inevitable next taxi. Similarly, animals whose lifestyles carry a high risk of accidental death are evolutionarily programmed to stint on repair and to age rapidly, even when living in the well-nourished safety of a laboratory cage. Mice, subject to high rates of predation in the wild, are evolutionarily programmed to invest less in repair and to age more rapidly than similar-sized caged birds that in the wild can escape predators by flying. Turtles, protected in the wild by a shell, are programmed to age more slowly than other reptiles, while porcupines, protected by quills, age more slowly than mammals comparable in size.

That generalization also fits us and our ape relatives. Ancient humans, who usually remained on the ground and defended themselves with spears and fire, were at lower risk of death from predators or from falling out of a tree than were arboreal apes. The legacy of the resultant evolutionary programming carries on today in that we live for several decades longer than do zoo apes living under comparable conditions of safety, health, and affluence. We must have evolved better repair mechanisms and decreased rates of senescence in the last seven million years, since we parted company from our ape relatives, came down out of the trees, and armed ourselves with spears and stones and fire.

Similar reasoning is relevant to our painful experience that everything in our bodies begins to fall apart as we grow older. Alas, that sad truth of evolutionary design is cost-efficient. You would be wasting biosynthetic energy, which otherwise could go into making babies, if you kept one part of your body in such great repair that it outlasted all your other parts and your resultant expected life span. The most efficiently constructed body is the one in which all organs wear out at approximately the same time.

The same principle, of course, applies to human-built machines, as illustrated in a story about that genius of cost-efficient automobile manufacture, Henry Ford. One day, Ford sent some of his employees to car junkyards, with instructions to examine the condition of the remaining parts in Model T Fords that had been junked. The employees brought back the apparently disappointing news that almost all components showed signs of wear. The sole exceptions were the kingpins, which remained virtually unworn. To the employees' surprise, Ford, instead of expressing pride in his well-made kingpins, declared that the kingpins were overbuilt, and that in the future they should be made more cheaply. Ford's conclusion may violate our ideal of pride in workmanship, but it made economic sense: he had indeed been wasting money on long-lasting kingpins that outlasted the cars in which they were installed.

The design of our bodies, which evolved through natural selection, fits Henry Ford's kingpin principle, with only one exception. Virtually every part of the human body wears out around the same time. The kingpin principle even fits men's reproductive tract, which undergoes no abrupt shutdown but does gradually accumulate a variety of problems, such as prostate hypertrophy and decreasing sperm count, to different degrees in different men. The kingpin principle also fits the bodies of animals. Animals caught in the wild show few signs of age-related deterioration because a wild animal is likely to die from a predator or accident when its body becomes significantly impaired. In zoos and laboratory cages, however, animals exhibit gradual age-related deterioration in every body part just as we do.

That sad message applies to the female as well as the male reproductive tract of animals. Female rhesus macaques run out of functional eggs around age thirty; fertilization of eggs in aged rabbits becomes less reliable; an increasing fraction of eggs are abnormal in aging hamsters, mice, and rabbits; fertilized embryos are increasingly unviable in aged hamsters and rabbits; and aging of the uterus itself leads to increasing embryonic mortality in hamsters, mice, and rabbits. Thus, the female reproductive tract of animals is a microcosm of the whole body in that everything that could go wrong with age may in fact go wrong— at different ages in different individuals.

The glaring exception to the kingpin principle is human female menopause. In all women within a short age span, it shuts down decades before expected death, even before the expected death of many hunter-gatherer women. It shuts down for a physiologically trivial reason—the exhaustion of functional eggs—that would have been easy to eliminate just by a mutation that slightly altered the rate at which eggs die or become unresponsive. Evidently, there was nothing physiologically inevitable about human female menopause, and there was nothing evolutionarily inevitable about it from the perspective of mammals in general. Instead, the human female, but not the human male, has become specifically programmed by natural selection, at some time within the last few million years, to shut down reproduction prematurely. That premature senescence is all the more surprising because it goes against an overwhelming trend: in other respects, we humans have evolved delayed rather than premature senescence.

•••••

Theorizing about the evolutionary basis of human female menopause must explain how a woman's apparently counterproductive evolutionary strategy of making fewer babies could actually result in her making more babies. Evidently, as a woman ages, she can do more to increase the number of people bearing her genes by devoting herself to her existing children, her potential grandchildren, and her other relatives than by producing yet another child.

The evolutionary chain of reasoning rests on several cruel facts. One is the human child's long period of parental dependence, longer than in any other animal species. A baby chimpanzee starts gathering its own food as it becomes weaned by its mother. It gathers the food mostly with its own hands. (Chimpanzee use of tools, such as fishing for termites with grass blades or cracking nuts with stones, is of great interest to human scientists but of only limited dietary significance to chimpanzees.) The baby chimpanzee also prepares its food with its own hands. But human hunter-gatherers acquire most of their food with tools, such as digging sticks, nets, spears, and baskets. Much human food is also prepared with tools (husked, pounded, cut up, et cetera) and then cooked in a fire. We do not protect ourselves against dangerous predators with our teeth and strong muscles, as do other prey animals, but, again, with our tools. Even to wield all those tools is completely beyond the manual dexterity of babies, and to make the tools is beyond the abilities of young children. Tool use and tool making are transmitted not just by imitation but by language, which takes over a decade for a child to master.

As a result, a human child in most societies does not become capable of economic independence or adult economic function until his or her teenage years or twenties. Until then, the child remains dependent on his or her parents, especially on the mother, because, as we saw in previous chapters, mothers tend to provide more child care than do fathers. Parents are important not only for gathering food and teaching tool making but also for providing protection and status within the tribe. In traditional societies, the early death of either the mother or the father prejudiced a child's life even if the surviving parent remarried, because of possible conflicts with the stepparent's genetic interests. A young orphan who was not adopted had even worse chances of surviving.

Hence a hunter-gatherer mother who already has several children risks losing some of her genetic investment in them if she does not survive until the youngest is at least a teenager. That one cruel fact underlying human female menopause becomes more ominous in the light of another cruel fact: the birth of each child immediately jeopardizes a mother's previous children because of the mother's risk of death in childbirth. In most other animal species, that risk is insignificant. For example, in one study encompassing 401 pregnant female rhesus macaques, only one died in childbirth. For humans in traditional societies, the risk was much higher and increased with age. Even in affluent, twentieth-century Western societies, the risk of dying in childbirth is seven times higher for a mother over the age of forty than for a twenty-year-old mother. But each new child puts the mother's life at risk not only because of the immediate risk of death in childbirth but also because of

the delayed risk of death related to exhaustion by lactation, carrying a young child, and working harder to feed
more mouths.

Yet another cruel fact is that infants of older mothers are
themselves increasingly unlikely to survive or be healthy
because of age-related increases in the risks of abortion,
stillbirth, low fetal weight, and genetic defects. For instance, the risk of a fetus carrying the genetic condition
known as Down's syndrome increases with the mother's
age, from one in two thousand births for a mother under
thirty, one in three hundred for a mother between the ages
of thirty-five and thirty-nine, and one in fifty for a forty-
three-year-old mother, to the grim odds of one in ten for a
mother in her late forties.

Thus, as a woman gets older, she is likely to have accumulated more children; she has also been caring for them
longer, so she is putting a bigger investment at risk with
each successive pregnancy. But her chances of dying in or
after childbirth, and the chances that the fetus or infant
will die or be damaged, also increase. In effect, the older
mother is taking on more risk for less potential gain. That's
one set of factors that would tend to favor human female
menopause and that would paradoxically result in a
woman ending up with more surviving children by giving
birth to fewer children. Natual selection has not programmed menopause into men because of three more cruel
facts: men never die in childbirth and rarely die while copulating, and they are less likely than mothers to exhaust
themselves caring for infants.

A hypothetically nonmenopausal old woman who died
in childbirth, or while caring for an infant, would thereby
be throwing away even more than her investment in her
previous children. That is because a woman's children
eventually begin producing children of their own, and
those children count as part of the woman's prior investment. Especially in traditional societies, a woman's sur-

vival is important not only to her children but also to her grandchildren.

That extended role of postmenopausal women has been explored by Kristen Hawkes, the anthropologist whose research on men's roles I discussed in chapter 5. Hawkes and her colleagues studied foraging by women of different ages among the Hadza hunter-gatherers of Tanzania. The women who devoted the most time to gathering food (especially roots, honey, and fruit) were postmenopausal women. Those hardworking Hadza grandmothers put in an impressive seven hours per day, compared to a mere three hours for teenagers and new brides and four and a half hours for married women with young children. As one might expect, foraging returns (measured in pounds of food gathered per hour) increased with age and experience, so that mature women achieved higher returns than teenagers, but, interestingly, the grandmothers' returns were still as high as those of women in their prime. The combination of more foraging hours and an unchanged foraging efficiency meant that the postmenopausal grandmothers brought in more food per day than any of the younger groups of women, even though their large harvests were greatly in excess of what was required to meet their own personal needs and they no longer had dependent young children to feed.

Hawkes and her colleagues observed that the Hadza grandmothers were sharing their excess food harvest with close relatives, such as their grandchildren and grown children. As a strategy for transforming food calories into pounds of baby, it would be more efficient for an older woman to donate the calories to grandchildren and grown children rather than to infants of her own (even if she still could give birth) because the older mother's fertility would be decreasing with age anyway, whereas her own children would be young adults at peak fertility. Naturally, this food-sharing argument does not constitute the sole reproductive contribution of postmenopausal women in traditional societies. A grandmother also baby-sits her grandchildren, thereby helping her adult children churn out more babies bearing the grandmother's genes. In addition, grandmothers lend their social status to their grandchildren, as to their children.

If one were playing God or Darwin and trying to decide whether to make older women undergo menopause or remain fertile, one would draw up a balance sheet, contrasting the benefits of menopause in one column with its costs in the other column. The costs of menopause are the potential children that a woman forgoes by undergoing menopause. The potential benefits include avoiding the increased risk of death due to childbirth and parenting at an advanced age, and gaining the benefit of improved survival for one's grandchildren and prior children. The sizes of those benefits depends on many details: How large is the risk of death in and after childbirth? How much does that risk increase with age? How large would the risk of death be at the same age even without children or the burden of parenting? How rapidly does fertility decrease with age before menopause? How rapidly would it continue to decrease in an aging woman who did not undergo menopause? All these factors are bound to differ between societies and are not easy to estimate. Hence anthropologists remain undecided whether the two considerations that I have discussed so far—investing in grandchildren and protecting one's prior investment in existing children—suffice to offset menopause's foreclosed option of further children and thus to explain the evolution of human female menopause.

· · · · ·

But there is still one more virtue of menopause, one that has received little attention. That is the importance of old people to their entire tribe in preliterate societies, which constituted every human society in the world from the time of human origins until the rise of writing in Mesopotamia around 3300 B.C. Textbooks of human genetics regularly assert that natural selection cannot weed out mutations tending to cause damaging effects of age in old people. Supposedly there can be no selection against such mutations because old people are said to be "postreproductive." I believe that such assertions overlook an essential fact that distinguishes humans from most animal species. No human, except a hermit, is ever truly postreproductive in the sense of being unable to benefit the survival and reproduction of other people bearing one's genes. Yes, I grant that if any orangutans lived long enough in the wild to become sterile, they would count as postreproductive, since orangutans other than mothers with one young offspring tend to be solitary. I also grant that the contributions of very old people to modern literate societies tend to decrease with age—a new phenomenon at the root of the enormous problems that old age now poses, both for the elderly themselves and for the rest of society. Today, we moderns get most of our information through writing, television, or radio. We find it impossible to conceive of the overwhelming importance of elderly people in preliterate societies as repositories of information and experience.

Here is an example of that role. In my field studies of bird ecology on New Guinea and adjacent Southwest Pacific islands, I live among people who traditionally had been without writing, depended on stone tools, and subsisted by farming and fishing supplemented by much hunting and gathering. I am constantly asking villagers to tell me the names of local species of birds, animals, and other plants in their local language, and to tell me what they know about each species. It turns out that New Guineans and Pacific islanders possess an enormous fund of traditional biological knowledge, including names for a thousand or more species, plus information about each species' habitat, behavior, ecology, and usefulness to humans. All that information is important because wild plants and animals traditionally furnished much of the people's food and all of their building materials, medicines, and decorations.

Again and again, when I ask a question about some rare bird, I find that only the older hunters know the answer, and eventually I ask a question that stumps even them. The hunters reply, "We have to ask the old man [or the old woman]." They then take me to a hut, inside of which is an old man or woman, often blind with cataracts, barely able to walk, toothless, and unable to eat any food that hasn't been prechewed by someone else. But that old person is the tribe's library. Because the society traditionally lacked writing, that old person knows much more about the local environment than anyone else and is the sole source of accurate knowledge about events that happened long ago. Out comes the rare bird's name, and a description of it.

That old person's accumulated experience is important for the whole tribe's survival. For instance, in 1976 I visited Rennell Island in the Solomon Archipelago, lying in the Southwest Pacific's cyclone belt. When I asked about consumption of fruits and seeds by birds, my Rennellese informants gave Rennell-language names for dozens of plant species, listed for each plant species all the bird and bat species that eat its fruit, and stated whether the fruit is edible for people. Those assessments of edibility were ranked in three categories: fruits that people never eat; fruits that people regularly eat; and fruits that people eat only in famine times, such as after—and here I kept hearing a Rennell term initially unfamiliar to me—after the *hungi kengi.* Those words proved to be the Rennell name for the most destructive cyclone to have hit the island in living memory—apparently around 1910, based on people's references to datable events of the European colonial administration. The hungi kengi blew down most of Rennell's forest, destroyed gardens, and drove people to the brink of starvation. Islanders survived by eating the fruits of wild plant species that normally were not eaten, but doing so required detailed knowledge about which plants were poisonous, which were not poisonous, and whether and how the poison could be removed by some technique of food preparation.

When I began pestering my middle-aged Rennellese informants with my questions about fruit edibility, I was brought into a hut. There, in the back of the hut, once my eyes had become accustomed to the dim light, was the inevitable, frail, very old woman, unable to walk without support. She was the last living person with direct experience of the plants found safe and nutritious to eat after the hungi kengi, until people's gardens began producing again. The old woman explained to me that she had been a child not quite of marriageable age at the time of the hungi kengi. Since my visit to Rennell was in 1976, and since the cyclone had struck sixty-six years before, around 1910, the woman was probably in her early eighties. Her survival after the 1910 cyclone had depended on information remembered by aged survivors of the last big cyclone before the hungi kengi. Now, the ability of her people to survive another cyclone would depend on her own memories, which fortunately were very detailed.

Such anecdotes could be multiplied indefinitely. Traditional human societies face frequent minor risks that threaten a few individuals, and they also face rare natural catastrophes or intertribal wars that threaten the lives of everybody in the society. But virtually everyone in a small traditional society is related to each other. Hence it is not only the case that old people in a traditional society are essential to the survival of their own children and grandchildren. They are also essential to the survival of the hundreds of people who share their genes.

Any human societies that included individuals old enough to remember the last event like a hungi kengi had a better chance of surviving than did societies without such old people. The old men were not at risk from childbirth or from the exhausting responsibilities of lactation and child care, so they did not evolve protection by menopause. But old women who did not undergo menopause tended to be eliminated from the human gene pool because they remained exposed to the risk of childbirth and the burden of child care. At times of crisis, such as a hungi kengi, the prior death of such an older woman also tended to eliminate all of her surviving relatives from the gene pool—a huge genetic price to pay for the dubious privilege of continuing to produce another baby or two against lengthening odds. That importance to society of the memories of old women is what I see as a major driving force behind the evolution of human female menopause.

·····

Of course, humans are not the only species that lives in groups of genetically related animals and whose survival depends on acquired knowledge transmitted culturally (that is, nongenetically) from one individual to another. For instance, we are coming to appreciate that whales are intelligent animals with complex social relationships and complex cultural traditions, such as the songs of humpback whales. Pilot whales, the other mammal species in which female menopause is well documented, are a prime example. Like traditional hunter-gatherer human societies, pilot whales live as "tribes" (termed pods) of 50 to 250 individuals. Genetic studies have shown that a pilot whale pod constitutes in effect a huge family, all of whose individuals are related to each other, because neither males nor females resettle from one pod to another. A substantial percentage of the adult female pilot whales in a pod are postmenopausal. While childbirth is unlikely to be as risky to pilot whales as it is to women, female menopause may have evolved in that species because nonmenopausal old females tended to succumb under the burdens of lactation and child care.

There are also other social animal species for which it remains to be established more precisely what percentage of females reach postmenopausal age under natural conditions. Those candidate species include chimpanzees, bonobos, African elephants, Asian elephants, and killer whales. Most of those species are now losing so many individuals to human depredations that we may already have lost our chance to discover whether female menopause is biologically significant for them in the wild. However, scientists have already begun to gather the relevant data for killer whales. Part of the reason for our fascination with killer whales and all of those other big social mammal species is that we can identify with them and their social relationships, which are similar to our own. For just that reason, I would not be surprised if some of those species too turn out to make more by making less.

....

CHAPTER 7

TRUTH IN ADVERTISING

The Evolution of Body Signals

Two friends of mine, a husband and wife whom I shall re-name Art and Judy Smith to preserve anonymity, had gone through a difficult time in their marriage. After both had a series of extramarital affairs, they had separated. Recently, they had come back together, in part because the separation had been hard on their children. Now Art and Judy were working to repair their damaged relationship, and both had promised not to resume their infidelities, but the legacy of suspicion and bitterness remained.

It was in that frame of mind that Art phoned home one morning while he was out of town on a business trip of a few days. A man's deep voice answered the phone. Art's throat choked instantly as his mind groped for an explanation. (*Did I dial the wrong number? What is a man doing there?*) Not knowing what to say, Art blurted out, "Is Mrs. Smith there?" The man answered matter-of-factly, "She's upstairs in the bedroom, getting dressed."

In a flash, rage swept over Art. He screamed inwardly to himself, "She's back to her affairs! Now she's having some bastard stay overnight in my bed! He even answers the phone!" Art had rapid visions of rushing home, killing his wife's lover, and smashing Judy's head into the wall. Still hardly able to believe his ears, he stammered into the tele-phone, "Who . . . is . . . this?"

The voice at the other end cracked, rose from the bari-tone range to a soprano, and answered, "Daddy, don't you recognize me?" It was Art and Judy's fourteen-year-old son, whose voice was changing. Art gasped again, in a mixture of relief, hysterical laughter, and sobbing.

Art's account of that phone call drove home for me how even we humans, the only rational animal species, are still held in the irrational thrall of animal-like behavioral pro-grams. A mere one-octave change in the pitch of a voice ut-tering half a dozen banal syllables caused the image conjured up by the speaker to flip from threatening rival to unthreatening child, and Art's mood to flip from murder-ous rage to paternal love. Other equally trivial cues spell the difference between our images of young and old, ugly and attractive, intimidating and weak. Art's story illus-trates the power of what zoologists term a signal: a cue that can be recognized very quickly and that may be insignifi-cant in itself, but which has come to denote a significant and complex set of biological attributes, such as sex, age, aggression, or relationship. Signals are essential to animal communication—that is, the process by which one animal alters the probability of another animal behaving in a way that may be adaptive to one or both individuals. Small sig-nals, which in themselves require little energy (such as ut-tering a few syllables at a low pitch), may release behaviors that require a lot of energy (such as risking one's life in an attempt to kill another individual).

Signals of humans and other animals have evolved through natural selection. For example, consider two indi-vidual animals of the same species, differing slightly in size and strength, facing each other over some resource that would benefit either individual. It would be advanta-geous to both individuals to exchange signals that accu-rately indicate their relative strength, and hence the likely outcome of a fight. By avoiding a fight, the weaker individual is spared the likelihood of injury or death, while the stronger individual saves energy and risk.

How do animal signals evolve? What do they actually convey? That is, are they wholly arbitrary, or do they possess any deeper meaning? What serves to ensure reliability and to minimize cheating? We shall now explore these questions about the body signals of humans, especially our signals related to sex. However, it is useful to begin with an overview of signals in other animal species, for which we can gain clearer insights through doing controlled experiments impossible to do on humans. As we shall see, zoologists have been able to gain insights into animal signals by means of standardized surgical modifications of animals' bodies. Some humans do ask plastic surgeons to modify their bodies, but the result does not constitute a well-controlled experiment.

. . . . .

Animals signal each other through many channels of communication. Among the most familiar to us are auditory signals, such as the territorial songs by which birds attract mates and announce possession to rivals, or the alarm calls by which birds warn each other of dangerous predators in the vicinity. Equally familiar to us are behavioral signals: dog lovers know that a dog with its ears, tail, and hair on the neck raised is aggressive, but a dog with its ears and tail lowered and neck hair flat is submissive or conciliatory. Olfactory signals are used by many mammals to mark a territory (as when a dog marks a fire hydrant with the odors in its urine) and by ants to mark a trail to a food source. Still other modalities, such as the electrical signals exchanged by electric fishes, are unfamiliar and imperceptible to us.

While these signals that I have just mentioned can be rapidly turned on and off, other signals are wired either permanently or for extended times into an animal's anatomy to convey various types of messages. An animal's sex is indicated by the male/female differences in plumage of many bird species or by the differences in head shape between male and female gorillas or orangutans. As discussed in chapter 4, females of many primate species advertise their time of ovulation by swollen, brightly colored skin on the buttocks or around the vagina. Sexually immature juveniles of most bird species differ in plumage from adults; sexually mature male gorillas acquire a saddle of silvery hairs on the back. Age is signaled more finely in Herring Gulls, which have distinct plumages as juveniles and at one, two, three, and four or more years of age.

Animal signals can be studied experimentally by creating a modified animal or dummy with altered signals. For instance, among individuals of the same sex, appeal to the opposite sex may depend on specific parts of the body, as is well known for humans. In an experiment demonstrating this point, the tails of male Long-Tailed Widowbirds, an African species in which the male's sixteen-inch tail was suspected of playing a role in attracting females, were lengthened or shortened. It turns out that a male whose tail is experimentally cut down to six inches attracts few mates, while a male with a tail extended to twenty-six inches by attaching an extra piece with glue attracts extra mates. A newly hatched Herring Gull chick pecks at the red spot on its parent's lower bill, thereby inducing the parent to vomit up half-digested stomach contents to feed the chick. Being pecked on the bill stimulates the parent to vomit, but seeing a red spot against a pale background on an elongated object stimulates the chick to peck. An artificial bill with a red dot receives four times as many pecks as a bill lacking the dot, while an artificial bill of any other color receives only half as many pecks as a red bill. As a final example, a European bird species called the Great Tit has a black stripe on the breast that serves as a signal of social status. Experiments with radio-controlled, motor-operated tit models placed at bird feeders show that live tits flying into the feeder retreat if and only if the model's stripe is wider than the intruder's stripe.

. . . . .

One has to wonder how on Earth animals evolved so that something seemingly so arbitrary as the length of a tail, the color of a spot on a bill, or the width of a black stripe produces such big behavioral responses. Why should a perfectly good Great Tit retreat from food just because it sees another bird with a slightly wider black stripe? What is it about a wide black stripe that implies intimidating strength? One would think that an otherwise inferior Great Tit with a gene for a wide stripe could thereby gain undeserved social status. Why doesn't such cheating become rampant and destroy the meaning of the signal?

These questions are still unresolved and much debated by zoologists, in part because the answers vary for different signals and different animal species. Let's consider these questions for body sexual signals—that is, structures on the body of one sex but not the opposite sex of the same species, and that are used as a signal to attract potential mates of the opposite sex or to impress rivals of the same sex. Three competing theories attempt to account for such sexual signals.

The first theory, put forward by the British geneticist Sir Ronald Fisher, is termed Fisher's runaway selection model. Human females, and females of all other animal species, face the dilemma of selecting a male with which to mate, preferably one bearing good genes that will be passed on to the female's offspring. That's a difficult task because, as every woman knows all too well, females have no direct way to assess the quality of a male's genes. Suppose that a female somehow became genetically programmed to be sexually attracted to males bearing a certain structure that gives the males some slight advantage at surviving compared to other males. Those males with the preferred structure would thereby gain an additional advantage: they would attract more females as mates and hence transmit their genes to more offspring. Females who preferred males with the structure would also gain an advantage: they would transmit the gene for the structure to their sons, who would in turn be preferred by other females.

A runaway process of selection would then ensue, favoring those males with genes for the structure in an exaggerated size and favoring those females with genes for an exaggerated preference for the structure. From generation to generation the structure would grow in size or conspicuousness until it lost its original slight beneficial effect on survival. For instance, a slightly longer tail might be useful for flying, but a peacock's gigantic tail is surely no use in flying. The evolutionary runaway process would halt only when further exaggeration of the trait would become detrimental for survival.

A second theory, proposed by the Israeli zoologist Amotz Zahavi, notes that many structures functioning as body sexual signals are so big or conspicuous that they must indeed be detrimental to their owner's survival. For instance, a peacock's or widowbird's tail not only doesn't help the bird survive but actually makes life more difficult. Having a heavy, long, broad tail makes it hard to slip through dense vegetation, take flight, keep flying, and thereby escape predators. Many sexual signals, like a bowerbird's golden crest, are big, bright, conspicuous structures that tend to attract a predator's attention. In addition, growing a big tail or crest is costly in that it uses up a lot of an animal's biosynthetic energy. As a result, argues Zahavi, any male that manages to survive despite such a costly handicap is in effect advertising to females that he must have terrific genes in other respects. When a female sees a male with that handicap, she is guaranteed that he is not cheating by carrying the gene for a big tail and being otherwise inferior. He would not have been able to afford to make the structure, and would not still be alive, unless he were truly superior.

One can immediately think of many human behaviors that surely conform to Zahavi's handicap theory of honest signals. While any man can boast to a woman that he is rich and therefore she should go to bed with him in the hopes of enticing him into marriage, he might be lying. Only when she sees him throwing away money on useless expensive jewelry and sports cars can she believe him. Again, some college students make a show of partying on the night before a big examination. In effect, they are saying: "Any jerk can get an A by studying, but I'm so smart that I can get an A despite the handicap of not studying."

The remaining theory of sexual signals, as formulated by the American zoologists Astrid Kodric-Brown and James Brown, is termed "truth in advertising." Like Zahavi and unlike Fisher, the Browns emphasize that costly body structures necessarily represent honest advertisements of quality, because an inferior animal could not afford the cost. In contrast to Zahavi, who views the costly structures as a handicap to survival, the Browns view them as either favoring survival or being closely linked to traits favoring survival. The costly structure is thus a doubly honest ad: only a superior animal can afford its cost, and it makes the animal even more superior.

For instance, the antlers of male deer represent a big investment of calcium, phosphate, and calories, yet they are grown and discarded each year. Only the most well-nourished males—ones that are mature, socially dominant, and free of parasites—can afford that investment. Hence a female deer can regard big antlers as an honest ad for male quality, just as a woman whose boyfriend buys and discards a Porsche sports car each year can believe his claim of being wealthy. But antlers carry a second message not shared with Porsches. Whereas a Porsche does not generate more wealth, big antlers do bring their owner access to the best pastures by enabling him to defeat rival males and fight off predators.

. . . . .

Let us now examine whether any of these three theories, devised to explain the evolution of animal signals, can also explain features of human bodies. But we first need to ask whether our bodies possess any such features requiring explanation. Our first inclination might be to assume that only stupid animals require genetically coded badges, like a red dot here and a black stripe there, in order to figure out each other's age, status, sex, genetic quality, and value as a potential mate. We, in contrast, have much bigger brains and far more reasoning ability than any other animal. Moreover, we are uniquely capable of speech and can thereby store and transmit far more detailed information than any other animal can. What need have we of red dots and black stripes when we routinely and accurately determine the age and status of other humans just by talking to them? What animal can tell another animal that it is twenty-seven years old, receives an annual salary of $125,000, and is second assistant vice president at the country's third largest bank? In selecting our mates and sex partners, don't we go through a dating phase that is in effect a long series of tests by which we accurately assess a prospective partner's parenting skills, relationship skills, and genes?

The answer is simple: nonsense! We too rely on signals as arbitrary as a widowbird's tail and a bowerbird's crest. Our signals include faces, smells, hair color, men's beards, and women's breasts. What makes those structures less ludicrous than a long tail as grounds for selecting a spouse— the most important person in our adult life, our economic and social partner, and the coparent of our children? If we think that we have a signaling system immune to cheating, why do so many people resort to makeup, hair dyes, and breast augmentation? As for our supposedly wise and careful selection process, all of us know that when we walk into a room full of unfamiliar people, we quickly sense who attracts us physically and who doesn't. That quick sense is based on "sex appeal," which just means the sum of the body signals to which we respond, largely unconsciously. Our divorce rate, now around 50 percent in the United States, shows that we ourselves acknowledge the failure of half of our efforts to select mates. Albatrosses and many other pair-bonded animal species have much lower "divorce" rates. So much for our wisdom and their stupidity!

In fact, like other animal species, we have evolved many body traits that signal age, sex, reproductive status, and individual quality, as well as programmed responses to those and other traits. Attainment of reproductive maturity is signaled in both human sexes by the growth of pubic and axillary hair. In human males it is further signaled by the growth of a beard and body hair and by a drop in the pitch of the voice. The episode with which I began this chapter illustrates that our responses to those signals can be as specific and dramatic as a gull chick's response to the red spot on its parent's bill. Human females additionally signal reproductive maturity by expansion of the breasts. Later in life, we signal our waning fertility and (in traditional societies) attainment of wise elder status by the whitening of our hair. We tend to respond to the sight of body muscles (in appropriate amounts and places) as a signal of male physical condition, and to the sight of body fat (also in appropriate amounts and places) as a signal of female physical condition. As for the body signals by which we select our mates and sex partners, they include all those same signals of reproductive maturity and physical condition, with variation among human populations in the signals that one sex possesses and that the other sex prefers.

For instance, men vary around the world in the luxuriance of their beard and body hair, while women vary geographically in the size and shape of their breasts and nipples and in their nipple color. All of these structures serve us humans as signals analogous to the red dots and black stripes of birds. In addition, just as women's breasts simultaneously perform a physiological function and serve as a signal, I shall consider later in this chapter whether the same might be true for men's penises.

•••••

Scientists seeking to understand the corresponding signals of animals can carry out experiments involving mechanical modifications of an animal's body, such as shortening a widowbird's tail or painting over a gull's red spot. Legal obstacles, moral compunctions, and ethical considerations prevent us from performing such controlled experiments on humans. Also preventing us from understanding human signals are our own strong feelings that cloud our objectivity about them, and the great degree of cultural variation and individually learned variation in both our preferences and our bodies' self-modifications. However, such variation and self-modification can also help us gain understanding by serving as natural experiments, albeit ones lacking experimental controls. At least three sets of human signals seem to me to conform to Kodric-Brown's and Brown's truth-in-advertising model: men's body muscle, facial "beauty" in both sexes, and women's body fat.

Men's body muscle tends to impress women as well as other men. While the extreme muscle development of professional bodybuilders strikes many people as grotesque, many (most?) women find a well-proportioned muscular man more attractive than a scrawny man. Men also use the muscular development of other men as a signal—for example, as a way of quickly assessing whether to get into a fight or to retreat. A typical example involves a magnificently muscular instructor named Andy at the gymnasium where my wife and I exercise. Whenever Andy lifts weights, the eyes of all the women and men in the gym are on him. When Andy explains to a customer how to use one of the gym's exercise machines, he begins by demonstrating the machine's operation himself while asking the customer to place a hand on the relevant muscle on Andy's body so that the customer can understand the correct motion. Undoubtedly, this means of explanation is pedagogically useful, but I am sure that Andy also enjoys the overwhelming impression that he leaves.

At least in traditional societies based on human muscle power rather than on machine power, muscles are a truthful signal of male quality, like a deer's antlers. On the one hand, muscles enable men to gather resources such as food, to construct resources such as houses, and to defeat rival men. In fact, muscles play a much larger role in a traditional man's life than do antlers in the life of a deer, which uses antlers only in fighting. On the other hand, men with other good qualities are better able to acquire all the protein required to grow and maintain big muscles. One can fake one's age by dyeing one's hair, but one cannot fake big muscles. Naturally, men did not evolve muscles solely to impress other men and women, in the way that male bowerbirds evolved a golden crest solely as a signal to impress other bowerbirds. Instead, muscles evolved to perform functions, and men and women then evolved or learned to respond to muscles as a truthful signal.

A beautiful face may be another truthful signal, although the underlying reason is not as transparent as in the case of muscles. If you stop to think about it, it may seem absurd that our sexual and social attractiveness depends on facial beauty to such an inordinate degree. One might reason that beauty says nothing about good genes, parenting qualities, or food-gathering skills. However, the face is the part of the body most sensitive to the ravages of age, disease, and injury. Especially in traditional societies, individuals with scarred or misshapen faces may thereby be advertising their proneness to disfiguring infections, inability to take care of themselves, or burden of parasitic worms. A beautiful face was thus a truthful signal of good health that could not be faked until twentieth-century plastic surgeons perfected facelifts.

Our remaining candidate for a truthful signal is women's body fat. Lactation and child care are a big energy drain on a mother, and lactation tends to fail in an undernourished mother. In traditional societies before the advent of infant formulas and before the domestication of milk-producing hoofed animals, a mother's lactational failure would have been fatal to her infant. Hence a woman's body fat would be a truthful signal to a man that she was capable of rearing his child. Naturally, men should prefer the correct amount of fat: too little could be a harbinger of lactational failure, but too much could signal difficulties in walking, poor food-gathering ability, or early death from diabetes.

Perhaps because fat would be difficult to discern if it were spread uniformly over the body, women's bodies have evolved with fat concentrated in certain parts that are readily visible and assessed, although the anatomical location of those fat deposits varies somewhat among human populations. Women of all populations tend to accumulate fat in the breasts and hips, to a degree that varies geographically. Women of the San population native to southern Africa (the so-called Bushmen and Hottentots) and women of the Andaman Islands in the Bay of Bengal accumulate fat in the buttocks, producing the condition known as steatopygia. Men throughout the world tend to be interested in women's breasts, hips, and buttocks, giving rise in modern societies to yet another surgical method of fake signals, breast enhancement. Of course, one can object that some individual men are less interested than other men in these signs of female nutritional status, and that the relative popularity of skinny and plump fashion models fluctuates from year to year as fads. Nevertheless, the overall trend in male interest is clear.

Suppose one were again playing God or Darwin and deciding where on a woman's body to concentrate body fat as a visible signal. The arms and legs would be excluded because of the resulting extra load on them during walking or use of the arms. That still leaves many parts of the torso where fat could be safely concentrated without impeding movement, and in fact I just mentioned that women of various populations have evolved three different signaling areas on the torso. Nevertheless, one has to ask whether the evolutionary choice of signaling area is completely arbitrary, and why there are no populations of women with other signaling locations, such as the belly or the middle of the back. Paired fat deposits on the belly would seem to create no more difficulties for locomotion than do our actual paired deposits in the breasts and buttocks. It is curious, however, that women of all populations have evolved fat deposition in the breasts, the organs whose lactational performance men may be attempting to assess by fat deposit signals. Hence some scientists have suggested that large fatty breasts are not only an honest signal of good overall nutrition but also a deceptive specific signal of high milk-producing ability (deceptive because milk is actually secreted by breast glandular tissue rather than by breast fat). Similarly, it has been suggested that fat deposition in the hips of women worldwide is also both an honest signal of good health and a deceptive specific signal suggesting a wide birth canal (deceptive because a truly wide birth canal would minimize the risk of birth traumas but mere fat hips would not).

. . . . .

At this point, I have to anticipate several objections to my assumption that the sexual ornamentation of women's bodies could have any evolutionary significance. Whatever the interpretation, it is of course a fact that women's bodies do possess structures functioning as sexual signals, and that men tend to be especially interested in those particular parts of women's bodies. In those respects women resemble females of other primate species living in troops that contain many adult males and adult females. Like humans, chimpanzees and baboons and macaques live in troops and have sexually ornamented females (as well as males). By contrast, female gibbons and the females of other primate species that live as solitary male-female pairs bear little or no sexual ornamentation. This correlation suggests that if and only if females compete intensively with other females for males' attention—for example, because multiple males and females encounter each other daily in the same troop—then females tend to evolve sexual ornamentation in an ongoing evolutionary contest to be more attractive. Females who do not have to compete on such a regular basis have less need of expensive body ornamentation.

In most animal species (including humans) the evolutionary significance of male sexual ornamentation is undisputed, because males surely compete for females. However, scientists have raised three objections to the interpretation that women compete for men and have evolved bodily ornaments for that purpose. First, in traditional societies at least 95 percent of women marry. This statistic seems to suggest that virtually any woman can get a husband, and that women have no need to compete. As one woman biologist expressed it to me, "Every garbage can has a lid, and there is usually a bad-looking man for every bad-looking woman."

But that interpretation is belied by all the effort that women consciously put into decoration and surgical modification of their bodies so as to be attractive. In fact, men vary greatly in their genes, in the resources that they control, in their parenting qualities, and in their devotion to their wives. Although virtually any woman can get some man to marry her, only a few women can succeed in getting one of the few high-quality men, for whom women must compete intensely. Every woman knows that, even though some male scientists evidently don't.

A second objection notes that men in traditional societies had no opportunity to choose their spouse, whether on the basis of sexual ornamentation or any other quality. Instead, marriages were arranged by clan relatives, who did the choosing, often with the motive of cementing political alliances. In reality, though, bride prices in traditional societies, such as the New Guinea societies where I work, vary according to a woman's desirability, the woman's health and probable mothering qualities being important considerations. That is, although a bridegroom's views about his bride's sex appeal may be ignored, his relatives who actually select the bride do not ignore their own views. In addition, men certainly consider a woman's sex appeal in selecting partners for extramarital sex, which is likely to account for a higher proportion of babies in traditional societies (where husbands don't get to follow their sexual preferences in selecting their wives) than in modern societies. Furthermore, remarriage following divorce or the death of the first spouse is very common in traditional societies, and men in those societies have more freedom in selecting their second spouse.

The remaining objection notes that culturally influenced beauty standards vary with time, and that individual men within the same society differ in their tastes. Skinny women may be out this year but in next year, and some men prefer skinny women every year. However, that fact is no more than noise slightly complicating but not invalidating the main conclusion: that men at all places and times have on the average preferred well-nourished women with beautiful faces.

· · · · ·

We have seen that several classes of human sexual signals—men's muscles, facial beauty, and women's body fat concentrated in certain places—apparently conform to the truth-in-advertising model. However, as I mentioned in discussing animals' signals, different signals may conform to different models. That's also true of humans. For example, the pubic and axillary hair that both men and women have evolved to grow in adolescence is a reliable but wholly arbitrary signal of attainment of reproductive maturity. Hair in those locations differs from muscles, beautiful faces, and body fat in that it carries no deeper message. It costs little to grow, and it makes no direct contribution to survival or to nursing babies. Poor nutrition may leave you with a scrawny body and disfigured face, but it rarely causes your pubic hair to fall out. Even weak ugly men and skinny ugly women sport axillary hair. Men's beards, body hair, and low-pitched voices as signals of adolescence, and men's and women's hair whitening as a signal of age, seem equally devoid of inner meaning. Like the red spot on a gull's bill and many other animal signals, these human signals are cheap and wholly arbitrary—many other signals can be imagined that would serve equally well.

Is there any human signal that exemplifies the operation of Fisher's runaway selection model or Zahavi's handicap principle? At first, we seem devoid of exaggerated signaling structures comparable to a widowbird's sixteen-inch tail. On reflection, however, I wonder whether we actually do sport one such structure: a man's penis. One might object that it serves a nonsignaling function and is nothing more than well-designed reproductive machinery. However, that is not a serious objection to my speculation: we have already seen that women's breasts simultaneously constitute signals and reproductive machinery. Comparisons with our ape relatives hint that the size of the human penis similarly exceeds bare functional requirements, and that that excess size may serve as a signal. The length of the erect penis is only about 1¼ inches in gorillas and 1½ inches in orangutans but 5 inches in humans, even though males of the two apes have much bigger bodies than men.

Are those extra couple of inches of the human penis a functionally unnecessary luxury? One counterinterpretation is that a large penis might somehow be useful in the wide variety of our copulatory positions compared to many other mammals. However, the 1½-inch penis of the male orangutan permits it to perform in a variety of positions that rival ours, and to outperform us by executing all those positions while hanging from a tree. As for the possible utility of a large penis in sustaining prolonged intercourse, orangutans top us in that regard too (mean duration fifteen minutes, versus a mere four minutes for the average American man).

A hint that the large human penis serves as some sort of signal may be gained by watching what happens when men take the opportunity to design their own penises, rather than remaining content with their evolutionary legacy. Men in the highlands of New Guinea do that by enclosing the penis in a decorative sheath called a phallocarp. The sheath is up to two feet long and four inches in diameter, often bright red or yellow in color, and variously decorated at the tip with fur, leaves, or a forked ornament. When I first encountered New Guinea men with phallocarps, among the Ketengban tribe in the Star Mountains last year, I had already heard a lot about them and was curious to see how they were used and how people explained them. It turned out that men wore their phallocarps constantly, at least whenever I encountered them. Each man owns several models, varying in size, ornamentation, and angle of erection, and each day he selects a model to wear according to his mood, much as each morning we select a shirt to wear. In response to my question as to why they wore phallocarps, the Ketengbans replied that they felt naked and immodest without them. That answer surprised me, with my Western perspective, because the Ketengbans were otherwise completely naked and left even their testes exposed.

In effect, the phallocarp is a conspicuous erect pseudo-penis representing what a man would like to be endowed with. The size of the penis that we evolved was unfortunately limited by the length of a woman's vagina. A phallocarp shows us what the human penis would look like if it were not subject to that practical constraint. It is a signal even bolder than the widowbird's tail. The actual penis, while more modest than a phallocarp, is immodestly large by the standards of our ape ancestors, although the chimpanzee penis has also become enlarged over the inferred ancestral state and rivals men's penises in size. Penis evolution evidently illustrates the operation of runaway selection just as Fisher postulated. Starting from a 1½-inch ancestral ape penis similar to the penis of a modern gorilla or orangutan, the human penis increased in length by a runaway process, conveying an advantage to its owner as an increasingly conspicuous signal of virility, until its length became limited by counterselection as difficulties fitting into a woman's vagina became imminent.

The human penis may also illustrate Zahavi's handicap model as a structure costly and detrimental to its owner. Granted, it is smaller and probably less costly than a peacock's tail. However, it is large enough that if the same quantity of tissue were instead devoted to extra cerebral cortex, that brainy redesigned man would gain a big advantage. Hence a large penis's cost should be regarded as a lost-opportunity cost: because any man's available biosynthetic energy is finite, the energy squandered on one structure comes at the expense of energy potentially available for another structure. In effect, a man is boasting, "I'm already so smart and superior that I don't need to devote more ounces of protoplasm to my brain, but I can instead afford the handicap of packing the ounces uselessly into my penis."

What remains debatable is the intended audience at which the penis's proclamation of virility is directed. Most men would assume that the ones who are impressed are women. However, women tend to report that they are more turned on by other features of a man, and that the sight of a penis is, if anything, unattractive. Instead, the ones really fascinated by the penis and its dimensions are men. In the showers in men's locker rooms, men routinely size up each other's endowment.

Even if some women are also impressed by the sight of a large penis or are satisfied by its stimulation of the clitoris and vagina during intercourse (as is very likely), it is not necessary for our discussion to degenerate into an either/or argument that assumes the signal to be directed at only one sex. Zoologists studying animals regularly discover that sexual ornaments serve a dual function: to attract potential mates of the opposite sex, and to establish dominance over rivals of the same sex. In that respect, as in many others, we humans still carry the legacy of hundreds of millions of years of vertebrate evolution engraved deeply into our sexuality. Over that legacy, our art, language, and culture have only recently added a veneer.

The possible signal function of the human penis, and the target of that signal (if there is one), thus remain unresolved questions. Hence this subject constitutes an appropriate ending to this book because it illustrates so well the book's main themes: the importance, fascination, and difficulties of an evolutionary approach to human sexuality. Penis function is not merely a physiological problem that can be straightforwardly cleared up by biomechanical experiments on hydraulic models, but an evolutionary problem as well. That evolutionary problem is posed by the fourfold expansion in human penis size beyond its inferred ancestral size over the course of the last 7 to 9 million years. Such an expansion cries out for a historical, functional interpretation. Just as we have seen with strictly female lactation, concealed ovulation, men's roles in society, and menopause, we have to ask what selective forces drove the historical expansion of the human penis and maintain its large size today.

Penis function is also an especially appropriate concluding subject because it seems at first so nonmysterious. Almost anyone would assert that the functions of the penis are to eject urine, inject sperm, and stimulate women physically during intercourse. But the comparative approach teaches us that those functions are accomplished elsewhere in the animal world by a relatively much smaller structure than the one with which we encumber ourselves. It also teaches us that such oversized structures evolve in several alternative ways that biologists are still struggling to understand. Thus, even the most familiar and seemingly most transparent piece of human sexual equipment surprises us with unsolved evolutionary questions.

For readers whose interest has been sufficiently aroused to read further, here are some suggestions. The first list consists of books on sexuality, behavior, primates, evolutionary reasoning, and related subjects. Many of them are written so as to be understandable to laypeople with no scientific training. They are available in large libraries, and many are still in print and available in bookstores. The second list consists of a dozen examples of technical articles, written for scientists and describing some of the specific studies that I discuss.

BOOKS

Alcock, John. *Animal Behavior: An Evolutionary Approach.* 5th ed. Sunderland, Mass.: Sinauer Associates, 1993.

Austin, C. R., and R. V. Short. *Reproduction in Mammals.* 2d ed., vols. 1–5. Cambridge: Cambridge University Press, 1982–86.

Chagnon, Napoleon A., and William Irons, eds. *Evolutionary Biology and Human Social Behavior: An Anthropological Perspective.* North Scituate, Mass.: Duxbury Press, 1979.

Cronin, Helena. *The Ant and the Peacock: Altruism and Sexual Selection from Darwin to Today.* Cambridge: Cambridge University Press, 1991.

Daly, Martin, and Margo Wilson. *Sex, Evolution, and Behavior.* 2d ed. Boston: Willard Grant Press, 1983.

Darwin, Charles. *The Descent of Man, and Selection in Rela-*

tion to Sex. London: Murray, 1871. Paperback reprint, Princeton, N.J.: Princeton University Press, 1981.

Diamond, Jared. *The Third Chimpanzee: The Evolution and Future of the Human Animal.* New York: HarperCollins, 1992.

Fedigan, Linda Marie. *Primate Paradigms: Sex Roles and Social Bonds.* Chicago: University of Chicago Press, 1992.

Goodall, Jane. *The Chimpanzees of Gombe: Patterns of Behavior.* Cambridge, Mass.: Harvard University Press, 1986.

Halliday, Tim. *Sexual Strategy.* Chicago: University of Chicago Press, 1980.

Hrdy, Sarah Blaffer. *The Woman That Never Evolved.* Cambridge, Mass.: Harvard University Press, 1981.

Kano, T. Takayoshi. *The Last Ape: Pygmy Chimpanzee Behavior and Ecology.* Stanford, Calif.: Stanford University Press, 1992.

Kevles, Bettyann. *Females of the Species: Sex and Survival in the Animal Kingdom.* Cambridge, Mass.: Harvard University Press, 1986.

Krebs, J. R., and N. B. Davies. *Behavioural Ecology: An Evolutionary Approach.* 3d ed. Oxford: Blackwell Scientific Publications, 1991.

Ricklefs, Robert E., and Caleb E. Finch. *Aging: A Natural History.* New York: Scientific American Library, 1995.

Rose, Michael R. *Evolutionary Biology of Aging.* New York: Oxford University Press, 1991.

Small, Meredith F. *Female Choices: Sexual Behavior of Female Primates.* Ithaca, N.Y.: Cornell University Press, 1993.

Smuts, Barbara B., Dorothy L. Cheney, Robert M. Seyfarth, Richard W. Wrangham, and Thomas T. Struhsaker, eds. *Primate Societies.* Chicago: University of Chicago Press, 1986.

Symons, Donald. *The Evolution of Human Sexuality.* New York: Oxford University Press, 1979.

Wilson, Edward O. *Sociobiology: The New Synthesis.* Cambridge, Mass.: Harvard University Press, 1975.

SCIENTIFIC ARTICLES
····

Alexander, Richard D. "How Did Humans Evolve?" Special publication no. 1. University of Michigan Museum of Zoology, Ann Arbor, 1990.

Emlen, Stephen T., Natalie J. Demong, and Douglas J. Emlen. "Experimental Induction of Infanticide in Female Wattled Jacanas." *Auk* 106 (1989): 1–7.

Francis, Charles M., Edythe L. P. Anthony, Jennifer A. Brunton, and Thomas H. Kunz. "Lactation in Male Fruit Bats." *Nature* 367 (1994): 691–92.

Gjershaug, Jan Ove, Torbjörn Järvi, and Eivin Røskaft. "Marriage Entrapment by 'Solitary' Mothers: A Study on Male Deception by Female Pied Flycatchers." *American Naturalist* 133 (1989): 273–76.

Greenblatt, Robert B. "Inappropriate Lactation in Men and Women." *Medical Aspects of Human Sexuality* 6, no. 6 (1972): 25–33.

Hawkes, Kristen. "Why Do Men Hunt? Benefits for Risky Choices." In *Risk and Uncertainty in Tribal and Peasant Economies,* edited by Elizabeth Cashdan (pp. 145–66). Boulder, Colo.: Westview Press, 1990.

Hawkes, Kristen, James F. O'Connell, and Nicholas G. Blurton Jones. "Hardworking Hadza Grandmothers." In *Comparative Socioecology: The Behavioral Ecology of Humans and Other Mammals,* edited by V. Standen and R. A. Foley (pp. 341–66). Oxford: Blackwell Scientific Publications, 1989.

Hill, Kim, and A. Magdalena Hurtado. "The Evolution of Premature Reproductive Senescence and Menopause in Human Females: An Evaluation of the 'Grandmother Hypothesis.'" *Human Nature* 2 (1991): 313–50.

Kodric-Brown, Astrid, and James H. Brown. "Truth in Advertising: The Kinds of Traits Favored by Sexual Selection." *American Naturalist* 124 (1984): 309–23.

Oring, Lewis W., David B. Lank, and Stephen J. Maxson. "Population Studies of the Polyandrous Spotted Sandpiper." *Auk* 100 (1983): 272–85.

Sillén-Tulberg, Birgitta, and Anders P. Møller. "The Relationship Between Concealed Ovulation and Mating Systems in Anthropoid Primates: A Phylogenetic Analysis." *American Naturalist* 141 (1993): 1–25.